FLOURISHING

FLOURISHING

*Health, Disease, and Bioethics
in Theological Perspective*

Neil Messer

WILLIAM B. EERDMANS PUBLISHING COMPANY
GRAND RAPIDS, MICHIGAN / CAMBRIDGE, U.K.

© 2013 Neil Messer
All rights reserved

Published 2013 by
Wm. B. Eerdmans Publishing Co.
2140 Oak Industrial Drive N.E., Grand Rapids, Michigan 49505 /
P.O. Box 163, Cambridge CB3 9PU U.K.
www.eerdmans.com

Printed in the United States of America

18 17 16 15 14 13   7 6 5 4 3 2 1

Library of Congress Cataloging-in-Publication Data

Messer, Neil.
Flourishing: health, disease, and bioethics in theological perspective / Neil Messer.
    pages   cm
Includes bibliographical references and index.
ISBN 978-0-8028-6899-2 (pbk.: alk. paper)
1. Health — Philosophy. 2. Health — Religious aspects. 3. Diseases —
Religious aspects. 4. Medical ethics. I. Title.

R723.M465   2013
174.2 — dc23
                                                        2013030578

The diagram on page 72 is from World Health Organization, *International Classification of Functioning, Disability, and Health,* Short Version. Geneva: World Health Organization, 2001, p. 26. Used with permission.

The excerpt from Dietrich Bonhoeffer's poem "Stations on the Road to Freedom" on page 133 is reprinted with the permission of Scribner Publishing Group from *Letters and Papers from Prison,* revised, enlarged ed., by Dietrich Bonhoeffer, translated from the German by R. H. Fuller, Frank Clark, et al. Copyright © 1953, 1967, 1971 by SCM Press Ltd. All rights reserved. The German is from Dietrich Bonhoeffer, *Widerstand und Ergebung.* © 1998, Gütersloher Verlagshaus, Gütersloh, in der Verlagsgruppe Random House GmbH.

To Fiona and Rebecca

Contents

ACKNOWLEDGMENTS ... viii

ABBREVIATIONS ... x

Introduction: The Importance of Concepts of Health and Disease ... xi

1. Philosophical Accounts of Health, Disease, and Illness ... 1
2. Disability Perspectives: Critical Insights and Questions ... 51
3. Theological Resources for Understanding Health and Disease ... 103
4. Theological Theses concerning Health, Disease, and Illness ... 163

Conclusion ... 201

BIBLIOGRAPHY ... 211

INDEX ... 227

Acknowledgments

In writing this book I have incurred many debts of gratitude, which I gladly acknowledge. Three of the four main chapters were written during three months' research leave from January through March 2012; I am indebted to the University of Winchester for supporting this leave, and to colleagues in the Department of Theology and Religious Studies and the Faculty of Humanities and Social Sciences who shouldered additional burdens as a result. During my leave I had the privilege of spending a highly enjoyable month as a Visiting Scholar at the Hastings Center in Garrison, New York, which proved a wonderfully congenial and productive place to think and write about health. Sincere thanks to everyone at the Center for their warm welcome, generous hospitality, and valuable advice — particularly the then President Tom Murray, Dan and Sidney Callahan, and Nancy Berlinger. Thank you also to my fellow Visiting Scholar Kieran Owens for many good conversations and some excellent vegetarian cooking.

Parts of the book have been presented as lectures and conference papers in various places, including the Society of Christian Ethics (San Jose, CA, January 2010, and Washington, DC, January 2012); the Theological Conference of St. Tikhon's Orthodox University, Moscow (November 2011); the Stead Center for Ethics and Values, Garrett-Evangelical Theological Seminary, Evanston (November 2012); the Philosophical Society of England (London, May 2012); and the Society for the Study of Christian Ethics (Cambridge, September 2012). I am very grateful for the invitations that led to some of these presentations and for the discussion from which I benefited on every occasion. Many thanks also to Simon Oliver for advice on teleology and for sharing some of his own unpublished work with me, and to Hans Reinders for kindly providing a pre-publication copy of a paper delivered to the Society for the Study of Christian Ethics in September 2012. Robert Song generously read a complete draft and made numerous valuable comments and suggestions, from which the final version has benefited greatly — though, of course, no one but me is responsible for the faults and weaknesses that remain. Sincere thanks to Jon Pott and his colleagues at Eerdmans for their enthusiasm for the book and their impressively efficient, supportive, and friendly management of its production.

Finally, heartfelt thanks as always to my wife Janet and my children Fiona and Rebecca, who cheerfully bore with my absence for a month and my preoccupation during other parts of last year — not to mention my antisocial habit of filling the house with books and papers — and whose love and presence sustain me in so many ways.

NEIL MESSER
University of Winchester
August 2013

Abbreviations

CD	Karl Barth, *Church Dogmatics*
DALY	Disability-adjusted life year
DSM	American Psychiatric Association, *Diagnostic and Statistical Manual of Mental Disorders*
HRQoL	Health-related quality of life
ICF	World Health Organization, *International Classification of Functioning, Disability, and Health*
ICIDH	World Health Organization, *International Classification of Impairment, Disability, and Handicap*
QALY	Quality-adjusted life year
QoL	Quality of life
SCG	Thomas Aquinas, *Summa Contra Gentiles*
ST	Thomas Aquinas, *Summa Theologica*
WHO	World Health Organization

INTRODUCTION

The Importance of Concepts of Health and Disease

This book has the modest aim of answering a simple-seeming question and indicating what some of the practical consequences of the answer might be. The question is: From the perspective of a Christian theological tradition, what should we understand by "health," "disease," "illness," and related terms? It might seem a piece of typically academic self-indulgence to devote an entire book to defining words that we use almost every day, whose meanings we often assume are obvious. However, the *importance*, at any rate, of understanding what we mean by health, disease, and illness is not hard to recognize. Any aspect of health care practice is, more or less by definition, a practical outworking of some notion of what "health" means and how it should be fostered and protected. Therefore ethical reflection on that practice will frequently be informed by assumptions about the meanings of health, disease, and illness, regardless of whether those doing the reflecting are aware of it. Various examples can be given to illustrate the point.[1]

First, in discussing the ethics of human genetic modification, it is customary to distinguish between gene therapy and genetic enhancement, that is, between genetic interventions intended to treat disease and those intended to enhance human traits not associated with disease. This distinction, however, is contested: some authors deny that it is a real distinction, or at any rate one with any moral significance, while others accept

---

[1]. For an additional example, namely public policy on human embryonic stem cell research in the UK, see Almut Caspary, *In Good Health: Philosophical-Theological Analysis of the Concept of Health in Contemporary Medical Ethics* (Stuttgart: Franz Steiner, 2010), pp. 25-42.

versions of the distinction but draw different moral inferences from it.[2] If a morally significant distinction between therapy and enhancement is sustainable, it will require an understanding of what is meant by disease, and how we know it when we see it.

Second, arguments about health care resource allocation often depend, explicitly or implicitly, on theories of health and well-being. For example, Norman Daniels is well known for adapting John Rawls's theory of justice as fairness to argue that meeting health needs has a "special moral importance" because it is necessary in order to safeguard each individual's fair share of the "normal opportunity range" for members of his or her society.[3] In order to demonstrate this special importance, Daniels needs a concept of health as normal functioning. In his original formulation of the argument, he adopted Christopher Boorse's "biostatistical" theory (to be discussed in chapter 1), because its naturalistic character appeared to allow him to avoid making normative judgments in identifying normal functioning and departures therefrom.[4] In his more recent presentation, he has moved away from Boorse's strictly naturalistic model — partly in response to the kinds of criticism to be surveyed in chapter 1 — and now favors Jerome Wakefield's account, in which diseases and other disorders are understood as "harmful dysfunctions."[5] Daniels believes that this account allows him to retain the scientifically objective notions of dysfunction and normal function required by his argument for the moral importance of health. His approach to health care resource allocation therefore seems to

---

2. For accounts that accept versions of the distinction, while taking different ethical views of the acceptability of enhancement, see Ted Peters, *Playing God? Genetic Determinism and Human Freedom* (New York: Routledge, 1997), pp. 144-53, and Robert Song, *Human Genetics: Fabricating the Future* (London: Darton, Longman, and Todd, 2002), pp. 58-78. For skepticism about the distinction, see John Harris, *Enhancing Evolution: The Ethical Case for Making Better People* (Princeton: Princeton University Press, 2007), pp. 44-58.

3. Norman Daniels, *Just Health: Meeting Health Needs Fairly* (Cambridge: Cambridge University Press, 2008), pp. 29-78 (quotations at pp. 29, 43), developing the account initially proposed in *Just Health Care* (Cambridge: Cambridge University Press, 1985). In the more recent version he has also argued that the capability approach of Amartya Sen and Martha Nussbaum and the egalitarianism of Richard Arneson and others, both critical of Rawls's theory in some respects, nonetheless agree on the need to protect opportunity and therefore lend support to his argument for the special moral importance of health; *Just Health*, pp. 63-77.

4. Daniels, *Just Health Care*, pp. 19-35.

5. Jerome C. Wakefield, "The Concept of Mental Disorder: On the Boundary between Biological Facts and Social Values," *American Psychologist* 47.3 (1992): 373-388.

stand or fall by the possibility of identifying normal function objectively, as Wakefield and Boorse both attempt to do.

The contrasting, utilitarian theory of the "quality-adjusted life year" (QALY) requires its users to quantify health-related quality of life (HRQoL).[6] Any measure of HRQoL will, of course, depend on prior assumptions about what is meant by "health" and "quality of life."[7] To the extent that the QALY approach depends on particular accounts of "health" and related concepts it will, like Daniels's account, be as conceptually strong or weak as the accounts on which it depends. More broadly, judgments about quality of life (QoL) inform a range of other ethical arguments and decisions in health care, including arguments about abortion, neonatal care, advance directives, end-of-life care, and assisted dying. Assumptions about health and human flourishing that underpin the way we understand QoL are therefore likely to have an influence on a wide range of important areas of ethical decision-making in health care, even if that influence is often unnoticed and unexamined.

As I have already observed, we often think it is perfectly obvious what is meant by health and disease, and no doubt our common-sense concepts of them serve us well enough in everyday life. However, when we try to examine these concepts more closely, they turn out to be more elusive and harder to define than we might expect. Moreover, their slippery nature matters most at just the points where the most difficult and contested decisions have to be made: consider, for example, the controversies within the American Psychiatric Association in the early 1970s over the removal of homosexuality from the *Diagnostic and Statistical Manual of Mental Disorders (DSM)*.[8]

If there is a distinctive understanding of health and disease to be found within the Christian tradition, this understanding could be expected to

---

6. The QALY approach has a widespread influence. For example, it is used in the UK to inform the recommendations made by the National Institute for Health and Care Excellence (NICE) about the publicly-funded prescribing of new medical treatments: National Institute for Health and Clinical Excellence, *Guide to the Methods of Technology Appraisal* (London: NICE, 2008), pp. 38-39. Available online at http://www.nice.org.uk/media/B52/A7/TAMethodsGuideUpdatedJune2008.pdf.

7. For a comparison of various HRQoL measures, see Elisabeth Stahl et al., "Health-related quality of life, utility, and productivity outcomes instruments: ease of completion by subjects with COPD," *Health and Quality of Life Outcomes* 1 (2003). Available online at http://www.hqlo.com/content/1/1/18.

8. See Ronald Bayer, *Homosexuality and American Psychiatry: The Politics of Diagnosis* (New York: Basic Books, 1981).

have important implications for Christian approaches to health care ethics. Articulating a theological understanding of health and disease, therefore, seems a promising way to learn more about the distinctive perspectives of the Christian tradition on a range of issues in health care ethics. In this book, I attempt that task, from a standpoint broadly within a Reformed Christian tradition, in dialogue with a range of other accounts and perspectives. Health, disease, illness, and related concepts have received extensive attention from philosophers of medicine since the mid-twentieth century, and chapter 1 surveys a representative sample of that literature. The chapter begins with the famous (or notorious) definition of health in the Preamble to the World Health Organization (WHO) Constitution (1948): "a state of complete physical, mental, and social well-being and not merely the absence of disease or injury,"[9] and Boorse's contrasting biostatistical theory, in which health is the absence of disease and diseases are states that interfere with species-typical natural functions. These accounts represent opposite ends of the scale on many questions about health and disease, and I survey various theories that occupy intermediate positions, such as those of Lennart Nordenfelt and K. W. M. Fulford. An important contrasting account is offered by S. Kay Toombs's phenomenological analysis of illness, which carefully maps the different perspectives of doctor and patient, and raises the question whether a distinction can be drawn between the subjective experience of "illness" and an objective, scientifically measurable state of "disease." Another complementary account, important for the connections it offers with later chapters, is provided by Iain Law and Heather Widdows, who make use of the "capability approach" associated with Amartya Sen and Martha Nussbaum to conceptualize health. The later sections of the chapter set up an extended argument between Jerome Wakefield and Christopher Megone. Wakefield's "harmful dysfunction" model offers an extensive analysis of health and disorder in terms of evolutionary biology, while Megone develops a teleological account of health, disease, and illness informed by a reading of Aristotle. The dispute between them brings into focus a range of questions that are central to an understanding of health, disease, and illness. For example: what can modern biology contribute to that understanding? Should health be understood in terms of the ends or goals of human life, and if so, how? What is the place of judgments about the good in understanding health, disease and illness?

9. World Health Organization, *Basic Documents*, 46th ed. (Geneva: World Health Organization, 2007), p. 1. Available online at http://www.who.int/gb/bd/.

These are all questions that must be taken up (and, where necessary, reframed) in a theological account of health, but before I attempt that, I turn in chapter 2 to a different area of discussion, namely the study of disability, in order to identify the critical questions that might be raised for concepts of health and disease by the experience of disability and academic reflection thereon. The chapter begins with a discussion of two contrasting ways of understanding disability: the "social model," particularly in the forms prominent in British disability studies, and the kind of "sociomedical" model represented by the WHO's *International Classification of Functioning, Disability, and Health (ICF)*.[10] I then offer a range of critical engagements with aspects of the philosophical discussion of health surveyed in chapter 1, including questions about normality, disorder and disability, quality of life, phenomenological perspectives on health, the body and disability, Nussbaum's use of the capability approach in relation to disability, and the significance of Aristotelian teleology for disability. Out of these engagements I identify a range of critical questions that the study of disability poses for an account of health, disease, and illness.

Chapter 3 begins an explicitly theological engagement with these questions, mapping some of the key theological resources available for such an engagement and identifying four areas in particular. The first is the Christian practice of healing and care of the sick, and theological reflection on that practice. I criticize the understanding of health as "wholeness" frequently found in the literature on the ministry of healing, arguing that it is less well supported biblically and theologically than is often supposed. I also reflect on the questions raised by the healing ministry about the meaning of suffering, and on the sometimes uneasy relationship between the Christian ministry of healing and the professional practice of health care. The second theological resource surveyed in this chapter is the important account of health and sickness developed by Karl Barth in the course of his ethics of creation. Barth defines health as "the strength for life" — the God-given ability to answer God's summons and simply *be* a creature of this particular kind — and sickness as "the weakness opposed to this strength."[11] This enables him to sustain a dialectical view of sickness: it is unequivocally an evil, but it can also be a

---

10. World Health Organization, *International Classification of Functioning, Disability, and Health*, Short Version (Geneva: WHO, 2001).

11. Karl Barth, *Church Dogmatics*, ed. and trans. Geoffrey W. Bromiley and Thomas F. Torrance, 13 vols. (Edinburgh: T&T Clark, 1956-75), III.4, pp. 356, 372.

reminder of our mortality and can therefore direct our ultimate hope towards God's promises.

Barth's account, I suggest, is implicitly teleological, and my third theological resource draws out that teleological perspective and makes it explicit. That resource is the teleological understanding of human life found in the thought of Thomas Aquinas. This allows us to develop an account in which part of what we mean by "human being" is a being whose good consists in the fulfillment of certain goals and ends characteristic of that kind of creature. Such a teleological view can engage critically and productively with Christopher Megone's philosophical teleology, explored in chapter 1. However, a Thomist teleology, too, requires some critical reappraisal at various points — for example, the importance it attaches to rationality — in the light of the final resource I bring into play. This fourth resource is a sample of the extensive recent theological literature on disability, which can variously reinforce, intensify, and modify some of the critical questions arising from the "secular" disability literature surveyed in chapter 2.

In chapter 4 I attempt to draw the threads of the preceding chapters together into a constructive theological account of health, disease, and illness guided by the "Barthian Thomist" approach outlined in the previous chapter. The account is presented as a series of sixteen theses, each accompanied by an outline explanation and defense, grouped under four headings: humans as creatures; health and creaturely flourishing; disease, suffering, evil, and sin; and, finally, practical implications.

The account given is one in which humans are understood as creatures of a particular kind, whose existence as such reflects the loving purposes of a good Creator. The human good is understood teleologically, as the fulfillment of goals and ends that are given in the structure of a creaturely life of this particular kind: to be a creature of this kind is to be one whose good consists in the fulfillment of certain characteristic goals and ends. An understanding of what it is to live a fulfilled creaturely life of this kind cannot be read off an empirical inspection of human nature (be that biological, sociological, psychological, or any other), but depends first and foremost on God's revelation in Christ. However, with that starting point established, it can be highly hospitable to insights and understanding from many different sources — including for the purposes of this book the biomedical sciences, the philosophy of medicine, and disability studies — and can engage "irenically and polemically" (to paraphrase Barth) with all these discourses and more besides. Moreover, the emphasis on the limits of human reason and its vulnerability to distortion by human sin enables this account to be sensitive to many of the critical perspectives offered in earlier chapters, insofar as they draw attention to the distorting effects of ideology, prejudice, and injustice on conceptions of the human good.

In this account, health is understood as an important, albeit limited, aspect of the human creaturely good. Diseases and illnesses are understood theologically as evils threatening this good, and this understanding underpins a strong Christian mandate for the practice of medicine and health care. Yet (in Dietrich Bonhoeffer's language) health is a penultimate, not an ultimate good, which betrays us if we invest ultimate hopes in it. Among other things, this implies that the care of the sick requires a refusal to abandon those who cannot be cured. Christian communities in their common life have (or ought to have) the resources to care when we cannot cure, as Stanley Hauerwas puts it.[12] Their life together witnesses (or ought to witness) to a hope that extends beyond this finite life, to God's promised good future.

The book ends with a brief conclusion in which I revisit the examples with which I began this introduction, and suggest how the account of health and disease developed in the book might address those issues. Space precludes more than an outline discussion in the conclusion, but I hope that the approach developed here will resource fuller analyses of the practical issues in future. I am also acutely conscious of other limitations of this work. For all the complexity of the interdisciplinary engagement attempted here, some disciplinary perspectives are underrepresented: sociological perspectives in particular do put in an appearance in chapter 2, but the book lacks a comprehensive engagement with the sociology of health. However, one book can only contain so much interdisciplinarity, and my hope is that the encounters which *are* offered here allow for a fruitful theological understanding to be developed and articulated.

---

12. Cf. Stanley Hauerwas with Richard Bondi and David B. Burrell, *Truthfulness and Tragedy: Further Investigations into Christian Ethics* (Notre Dame: University of Notre Dame Press, 1977), p. 196.

CHAPTER 1

Philosophical Accounts of Health, Disease, and Illness

Introduction

The meanings of "health," "disease," and related concepts have been extensively debated in the philosophical literature over several decades.[1] These debates have at times become both lengthy and intricate. Even when they are not at their most accessible, it is nonetheless important for the purpose of this book to attend to them in a certain amount of depth, because they offer accounts and insights and raise issues with which the theological treatment being developed here must engage. It is the purpose of this chapter to survey a representative sample of philosophical discussion in enough detail to identify the key insights and questions that emerge.

   1. A note on terminology: numerous different terms and concepts are used in the literature, including "health," "well-being," "disease," "illness," "sickness," "disorder," and "malady," but confusingly (though perhaps predictably), terms are not always employed to do the same work by different authors. For example, "disease" is sometimes taken to mean an objective condition, describable clinically and scientifically, as opposed to "illness," which refers to the subjective experience of being in less than full health. But the disease-illness distinction is not always set up in this way, and some accounts (as we shall see) suggest that it should not be made at all. Or again, not all deficits of health are diseases: some, for instance, are injuries, which are commonly distinguished from diseases in both ordinary and professional speech. For this reason, some authors use terms such as "disorder" or "malady" to denote a broader class of health deficits than diseases alone. In what follows, "health," "disease," and "illness" will be the terms used most frequently, though when the occasion demands, "disorder" will be used as a broader category that includes, but is not limited to, disease. Finally, "the health concepts" is sometimes used (for example by Fulford, below, pp. 16-18) as a collective shorthand for all these health-related terms. For convenience, this usage will from time to time be adopted in what follows.

It will be helpful to keep in mind some of the main questions in play in these debates, questions that will emerge at intervals during the following survey of the various accounts. In particular:

1. How narrow or broad a range of human goods is included in "health" and related concepts?

2. What is the relationship between health and other human goods?

3. Which aspects or components of the health concepts are: (a) value-free, and which value-laden; (b) objective, and which subjective; (c) universally human, and which relative to individuals' lives and circumstances or to social and cultural contexts? These oppositions are related, and are sometimes conflated in discussions of health, but they are distinct. For example, it often seems to be taken for granted that if an account of health incorporates evaluative judgments about good or harm, it will also be subjective and relative to context; but Martha Nussbaum's critique of Amartya Sen's "capability" approach, quoted below,[2] suggests the possibility of an account of health, rooted in an Aristotelian notion of human flourishing, that would be "value"-laden, yet also objective and universally human in at least some of its elements.

4. Should health be understood teleologically, and if so, what will a teleological account entail?

The World Health Organization Definition[3]

Discussions of health and disease frequently begin with the well-known definition of health in the Preamble to the World Health Organization (WHO) Constitution of 1948: "a state of complete physical, mental, and social well-being and not merely the absence of disease or infirmity."[4] The Preamble claims that the "enjoyment of the highest attainable state of health" is a fundamental human right and that "the health of all peoples is fundamental to peace and security."[5] As Daniel Callahan observes, the historical context in which the WHO Constitution was drafted helps to explain the grandiose vi-

---

2. See below, p. 26.
3. Parts of this and the next sections are drawn from Neil Messer, "Toward a Theological Understanding of Health and Disease," *Journal of the Society of Christian Ethics* 31.1 (2011): 161-178.
4. World Health Organization, *Basic Documents*, 46th ed. (Geneva: World Health Organization, 2007), p. 1. Available online at http://www.who.int/gb/bd/.
5. WHO, *Basic Documents*, p. 1.

sion expressed in this definition.[6] The WHO was formed in the aftermath of the Second World War, in the context of "a conviction that health was intimately related to economic and cultural welfare"[7] (which was in turn essential for the preservation of peace), and a powerful confidence in the potential of scientific medicine. A statement of Dr. Brock Chisholm, the first director of the WHO, captures the mood nicely:

> The world is sick and the ills are due to the perversion of man; his inability to live with himself. The microbe is not the enemy; science is sufficiently advanced to cope with it were it not for the barriers of superstition, ignorance, religious intolerance, misery, and poverty. . . . These psychological evils must be understood in order that a remedy might be prescribed, and the scope of the task before the Committee [charged with drafting the WHO Constitution] therefore knows no bounds.[8]

The WHO definition captures some important insights and intuitions. It gives an account of why health *matters* for human well-being: as Callahan puts it, "the intimate connection between the good of the body and the good of the self, not only the individual self but the social community of selves."[9] It articulates the insight that health is a necessary condition for the realization of many other human goods and goals, and the intuition that it is not fully described by the absence of disease states or the deviation of biological functions from statistical norms.

However, as an attempt to define health in a way that can inform practice and moral discernment in health care, it has some serious — and by now well-known — problems and drawbacks. One is its vagueness: the denial that health can be defined without remainder as the absence of disease requires the introduction of another term by which to define it. The term chosen by the WHO is "well-being," but this is if anything even less well-defined and more contested than "health," so it offers little by way of conceptual clarification.[10] Another problem is the idealizing character of the

---

6. Daniel Callahan, "The WHO Definition of 'Health,'" *Hastings Center Studies* 1.3 (1973): 77-88 (pp. 79-80).
7. Callahan, "The WHO Definition," p. 79.
8. Quoted by Callahan, "The WHO Definition," pp. 79-80.
9. Callahan, "The WHO Definition," p. 86.
10. Iain Law and Heather Widdows, "Conceptualizing Health: Insights from the Capability Approach," *Health Care Analysis* 16 (2008): 303-314 (p. 304).

definition: by identifying health with *complete* well-being, it tends to make the pursuit of health a never-ending quest for an unattainable goal, and the assertion of a human right to "the highest attainable state of health," so defined, could turn out to be so over-ambitious as to be vacuous.[11]

A third problem with the WHO definition is its all-embracing character, which if pushed to its conclusion would tend to medicalize every area of human life. Callahan points out that by assimilating all human well-being to the concept of health, the WHO definition "by implication ... makes the medical profession the gatekeeper for happiness and social well-being,"[12] potentially placing awesome power in the hands of health care professionals and a weighty responsibility on their shoulders. Two contrasting but interrelated dangers follow from this.[13] First, if all suffering and disorder (including social disorder) is classified as a problem of health, it comes to seem plausible to place anyone who contributes to such disorder (which probably means all of us, to a greater or lesser extent) in a sociological "sick role," in which one is not held morally responsible for behavior normally considered reprehensible. Therefore, if the language of health and sickness is stretched to include social and communal disorder, it may become difficult to speak of our own or others' moral responsibility for the well-being of our community and world. The second danger is the other side of the same coin: if health is understood in such a wide-ranging way, it becomes crucial for the survival and well-being of a society that all of its members are "healthy" in this comprehensive sense. A "tyranny of health" could then develop in which health becomes a moral and societal obligation, and a society through its health professionals adopts highly coercive policies towards those judged to be "sick." Callahan considers such re-description of social deviance or misbehavior as forms of sickness to be a contributory factor in the abuses of psychiatry and the over-prescribing of psychiatric drugs long criticized by "anti-psychiatrists" such as Thomas Szasz.[14]

Furthermore, if the concept of health is made into a theory of everything, an all-embracing account of every aspect of human well-being, then

---

11. Cf. John D. Arras and Elizabeth M. Fenton, "Bioethics and Human Rights: Access to Health-Related Goods," *Hastings Center Report* 39.5 (2009): 27-38.

12. Callahan, "The WHO Definition," p. 81.

13. Callahan, "The WHO Definition," pp. 82-83.

14. See, e.g., Thomas S. Szasz, *The Myth of Mental Illness: Foundations of a Theory of Personal Conduct*, rev. ed. (New York: Harper and Row, 1974); "Mental Illness Is Still a Myth," *Society* 31.4 (1994): 34-39; "Second Commentary on 'Aristotle's Function Argument,'" *Philosophy, Psychiatry, and Psychology* 7.1 (2000): 3-16.

it becomes difficult if not impossible to give an account of a diversity of human goods, goods that in a finite and flawed world might sometimes be in conflict with each other and might need to be ranked in importance. There are many human endeavors, such as caring for needy and suffering people, the struggle for political justice, artistic expression, athletic excellence, and even academic research, which have been considered by some of their practitioners to be worth pursuing even at a heavy cost to physical health, emotional well-being, or personal relationships. It is beside the point whether those athletes who have risked injury for the sake of success, those priests and pastors whose ministries in conditions of squalor and disease have brought them to an early grave, those political activists who have risked liberty, relationships, and sometimes their own lives for their causes, or those martyrs who have counted faithful Christian witness a more important good than the preservation of their bodily life were right to do so.[15] The point is simply that there are such choices, and an account of health which tends to obscure them or render them unintelligible must be considered questionable. While it could be objected that the WHO definition does not make these choices unintelligible because the conflict and ranking of different human goods can be understood as the conflict and ranking of different aspects of "health," we might then ask what is gained by calling all of this "health." If "health" is taken to mean everything, it could end up meaning nothing.

A final difficulty with conceptualizing health as a theory of everything is that, while the meaning of "health" is stretched too wide, we are also left with an inadequate notion of "everything." In other words, the WHO definition not only suggests too wide-ranging a concept of health, but also implies too narrow an account of total human well-being or flourishing. In particular, it has often been criticized for omitting spiritual well-being (to use a rather ill-

15. See further Joseph Boyle, "Limiting Access to Health Care: A Traditional Roman Catholic Analysis," in *Allocating Scarce Medical Resources: Roman Catholic Perspectives*, ed. H. Tristram Engelhardt Jr. and Mark J. Cherry (Washington, DC: Georgetown University Press, 2002), pp. 79-80; Stephen E. Lammers and Allen Verhey, eds., *On Moral Medicine: Theological Perspectives in Medical Ethics*, 2nd ed. (Grand Rapids: Eerdmans, 1998), pp. 239-241; Neil Messer, "The Human Genome Project, Health, and the 'Tyranny of Normality,'" in *Brave New World? Theology, Ethics, and the Human Genome*, ed. Celia Deane-Drummond (London: T&T Clark, 2003), pp. 91-115. On one group of clergy well-known for their costly ministries in conditions of urban squalor and poverty, see, e.g., John Shelton Reed, "'Ritualism Rampant in East London': Anglo-Catholicism and the Urban Poor," *Victorian Studies* 31.3 (1988): 375-403. For one family's story of political activism and its personal cost, see Gillian Slovo, *Every Secret Thing: My Family, My Country* (London: Little, Brown, 1997).

adapted term) from the picture, notwithstanding later attempts to include spiritual well-being;[16] and indeed, it is hardly surprising if WHO policy documents *do* fail to give an adequate account of this aspect of human life.

The Biostatistical Model: Christopher Boorse

At the other end of the scale from the WHO definition, and partly in reaction to it, lies the kind of "biostatistical model" most famously associated with Christopher Boorse. In his early work on health Boorse, like Thomas Szasz, has the burgeoning social and political role of psychiatry in his sights.[17] As Bill Fulford notes, Boorse aims to place medical theory on as solid a scientific foundation as possible, drawing a clear distinction between medical practice (which is avowedly and properly value-laden) and medical theory (which should be scientific, objective, and value-free).[18] Boorse attempts to secure the objective, value-free character of medical theory by defining health and disease in terms of statistical normality:

> An organism is *healthy* at any moment in proportion as it is not diseased; and a *disease* is a type of internal state of the organism which:
> (i) interferes with the performance of some natural function — i.e., some species-typical contribution to survival and reproduction, characteristic of the organism's age; and
> (ii) is not simply in the nature of the species, i.e., is either atypical of the species, or, if typical, mainly due to environmental causes.[19]

In his early papers, he maintains the distinction between value-free medical theory and value-laden medical practice by means of a distinction between "disease" and "illness," the additional conditions for the latter being

---

16. John Wilkinson, *The Bible and Healing: A Medical and Theological Commentary* (Grand Rapids: Eerdmans, 1998), p. 19.

17. Christopher Boorse, "On the Distinction Between Disease and Illness," *Philosophy and Public Affairs* 5.1 (1975): 49-68.

18. K. W. M. Fulford, "'What Is (Mental) Disease?': An Open Letter to Christopher Boorse," *Journal of Medical Ethics* 27 (2001): 80-85.

19. Christopher Boorse, "What a Theory of Mental Health Should Be," *Journal for the Theory of Social Behavior* 6 (1976): 61-84 (pp. 62-63, emphasis original). This biostatistical definition of health and disease is developed and defended more fully in Christopher Boorse, "Health as a Theoretical Concept," *Philosophy of Science* 44.4 (1977): 542-573.

that it is "(i) undesirable for its bearer; (ii) a title to special treatment; and (iii) a valid excuse for normally criticizable behavior."[20] In a 1997 defense of his theory, Boorse retracts this distinction between disease and illness, while continuing to distinguish on more complex grounds between (value-free) medical theory and (value-laden) medical practice.[21] Others, however, do maintain various forms of a disease/illness distinction, as we shall see.

In contrast to the WHO definition, which is both all-embracing and avowedly normative, Boorse's biostatistical theory does at least four important things. First, it conceptualizes the domain of health as a particular, limited domain of human experience, and represents an attempt to define the boundaries of that domain as clearly as possible. Second, it offers a way of connecting an understanding of health with the theoretical framework of the biological sciences, namely evolutionary biology. This connection is implicit in the phrase "species-typical contribution to survival and reproduction," and is drawn out more fully and explicitly by others such as Jerome Wakefield, as will be discussed below.[22] Third, in his initial disease-illness distinction and in his later revision of it, Boorse offers one kind of answer to the following questions: which components of an account of health are value-free, which are value-laden, and how are the value-free and value-laden components related to one another? Fourth, it offers one answer to the question, which of the health concepts has conceptual priority? For Boorse, it is "disease," since "health" is defined in opposition to "disease" and "illness" as a subset of it. These four issues — the limits of the concept of "health," its connection with the biological sciences, the location and limits of value within an account of health, and the conceptual hierarchy of the health concepts — have all been preoccupations of much subsequent philosophical discussion, as will become evident through the present chapter.

Over the years since it was first proposed, much ink has been spilled over the merits of Boorse's theory; indeed, as George Khushf recently observed, it remains the starting point for most current discussion of the concepts of health and disease.[23] It has received at least its fair share of cri-

---

20. Boorse, "On the Distinction Between Disease and Illness," p. 61.
21. Christopher Boorse, "A Rebuttal on Health," in *What Is Disease?* ed. James M. Humber and Robert F. Almeder (Totowa, NJ: Humana Press, 1997), pp. 1-134 (pp. 11-13).
22. See also Boorse, "Health as a Theoretical Concept," pp. 556-57.
23. George Khushf, "An Agenda for Future Debate on Concepts of Health and Disease," *Medicine, Health Care, and Philosophy* 10 (2007): 19-27 (p. 19). It must be said, however, that Khushf finds this state of affairs unsatisfactory, and proposes that Lennart Nordenfelt's rival theory (discussed below) serve as the starting point for future debate.

tique, along with some vigorous defense by Boorse himself and others. One criticism is that his definition is *too* narrow: that the use of "species-typical contribution to survival and reproduction" as the criterion by which to define the kind of loss of function that counts as a disease excludes some well-known and uncontroversial examples. Anne Gammelgaard, for example, identifies incontinence as a condition that "is disabling because of its social consequences rather than any detrimental consequences in relation to survival and reproduction."[24] Scott DeVito argues in similar vein that "life and reproduction by themselves do not specify *every* type of body-state that interests people or that falls under the category of diseases,"[25] citing blindness, deafness, multiple sclerosis, and myocardial infarction as examples of conditions that are counted as diseases partly on the grounds of their effect on quality of life, not only on survival and reproduction. This objection of Gammelgaard and DeVito, however, does not seem entirely successful, because Boorse's definition does not require that a disease actually *has* detrimental consequences for survival and reproduction, merely that it interrupts a *function* that contributes to survival and reproduction. Such loss of function need not *itself* threaten survival or reproduction to be counted a disease.[26]

Other objections to the theory focus in various ways on the concept of "natural function." For example, it has been argued that the notion of a species-typical function is not sustainable.[27] A related objection questions whether evolutionary theory can support the concept of "species-typical contribution to survival and reproduction" on which Boorse's theory depends.[28] Since other theories of health, notably Wakefield's, make more explicit and extensive use of evolutionary theory to ground a concept of

24. Anne Gammelgaard, "Evolutionary Biology and the Concept of Disease," *Medicine, Health Care, and Philosophy* 3 (2000): 109-116 (p. 110).

25. Scott DeVito, "On the Value-Neutrality of the Concepts of Health and Disease: Unto the Breach Again," *Journal of Medicine and Philosophy* 25.5 (2000): 539-567 (p. 545).

26. For instance, in relation to Gammelgaard's example of incontinence, it is not hard to see how functions for controlling the elimination of bodily waste could have a survival value for members of the human species (for example by limiting the spread of infectious diseases) even if some kinds of failure in that function do not in themselves threaten survival or reproduction.

27. Ron Amundson, "Against Normal Function," *Studies in History and Philosophy of Biological and Biomedical Sciences* 31.1 (2000): 33-53. For Boorse's rejoinder, see Christopher Boorse, "Disability and Medical Theory," in *Philosophical Reflections on Disability*, Philosophy and Medicine 104, ed. D. C. Ralston and J. Ho (New York: Springer, 2010), pp. 55-88 (pp. 76-77).

28. See Gammelgaard, "Evolutionary Biology and the Concept of Disease."

biological function, objections concerned with the evolutionary basis of "natural function" will be discussed below in connection with Wakefield.

Probably the most familiar line of critique, however, is that Boorse does not succeed in expelling value from the concept of health. Fulford, for example, points out that in Boorse's own papers, he frequently slips into evaluative language. From the perspective of linguistic analysis, Fulford considers such slips significant: the evaluative language Boorse cannot help using gives a more reliable indication of the meaning of the concepts of health and disease than the formal definitions he attempts to defend; so, Fulford concludes, Boorse's own "naturalist" account does not work.[29] Neither would a "descriptivist" theory in which value-laden terms are translatable into value-free terms.[30] According to Fulford, the reason that a descriptivist theory of health cannot rescue Boorse is that descriptivism only works when the translation from value-laden to value-free commands a wide consensus. This condition is not satisfied in the case of talk about health and disease, because often there is no general agreement as to what purely descriptive state of affairs counts as "better" evaluatively, particularly in relation to *mental* health and disease.[31] This lack of consensus could of course invite the Szasz-inspired rejoinder that if so, we should jettison the concepts of mental health and disease and restrict ourselves to cases of physical health and disease where the translation into value-neutral descriptions can be made.[32] However, it is not clear that this kind of retrenchment would save Boorse's theory either, because it is doubtful that physical disease is the unproblematically value-free paradigm case that Szasz takes it to be.[33]

Fulford is by no means alone in making this critique: numerous other

---

29. K. W. M. Fulford, *Moral Theory and Medical Practice* (Cambridge: Cambridge University Press, 1989), pp. 35-56; K. W. M. Fulford, "Teleology without Tears: Naturalism, Neo-Naturalism, and Evaluationism in the Analysis of Function Statements in Biology (and a Bet on the Twenty-first Century)," *Philosophy, Psychiatry, and Psychology* 7.1 (2000): 77-94; Fulford, "What Is (Mental) Disease?"

30. An oft-cited example is that the evaluative term "better," used of a share price on the stock market, is said to be translatable into the value-free term "higher": Fulford, "What Is (Mental) Disease?" p. 82.

31. Fulford, "What Is (Mental) Disease?" pp. 81-82.

32. See, e.g., Szasz, "Second Commentary on 'Aristotle's Function Argument.'"

33. Fulford, *Moral Theory and Medical Practice*, pp. 3-24; Christopher Megone, "Aristotle's Function Argument and the Concept of Mental Illness," *Philosophy, Psychiatry, and Psychology* 5.3 (1998): 187-201 (p. 188); Christopher Megone, "Mental Illness, Human Function, and Values," *Philosophy, Psychiatry, and Psychology* 7.1 (2000): 45-65 (pp. 47-49).

commentators have made similar points on a variety of grounds.[34] Scott DeVito, for example, argues that "[t]he reason medical researchers . . . study the body-states they do is that the explanation for the occurrence of those body-states is of value." It is *practical* knowledge, sought because it "enables medical practitioners to prevent the occurrence of disvalued body-states and to facilitate the occurrence of valued body-states."[35] Therefore if survival and reproduction are the biological goals that are of particular interest to biomedical scientists, it is not for value-free reasons, but because those goals are valued, and the value attributed to them motivates scientific efforts to learn how to promote them.

A related, more general concern is that Boorse, like many of the authors to be reviewed in this chapter, relies on a sharp separation of fact from value. This separation is often simply assumed by modern authors, but has been challenged by some commentators. Alasdair MacIntyre, for example, argues that it is a peculiarly modern distinction, characteristic of what he calls the "Enlightenment project."[36] Before the eighteenth century, it would not have seemed nearly so obvious that fact and value, or description and evaluation, should be sharply separated: to describe someone as, for example, a human being just was to describe them as a being whose good consisted in the fulfillment of certain goals. The fact-value separation has more recently also been challenged by Christopher Megone in proposing his Aristotelian account of health and defending it against critics such as Szasz and Wakefield.[37]

"Reverse Theories": Nordenfelt and Fulford

Boorse's biostatistical theory, as we have seen, gives priority to the concept of disease, defining health and illness in terms of disease. Other accounts, notably those of Lennart Nordenfelt and Bill Fulford, are sometimes described as "reverse theories," in the sense that one of the other elements is

---

34. In addition to Fulford, DeVito, and Khushf, see also, e.g., Richard M. Hare, "Health," *Journal of Medical Ethics* 12 (1986): 174-181.

35. DeVito, "On the Value-Neutrality of the Concepts of Health and Disease," p. 544.

36. Alasdair MacIntyre, *After Virtue: A Study in Moral Theory*, 2nd ed. (London: Duckworth, 1985), pp. 56-59. See, further, Neil Messer, *Selfish Genes and Christian Ethics: Theological and Ethical Reflections on Evolutionary Biology* (London: SCM, 2007), p. 105.

37. Megone, "Aristotle's Function Argument"; "Mental Illness, Human Function, and Values"; see further below, pp. 38-48.

given conceptual priority and disease is defined in terms of that element.[38] For Nordenfelt, the prior concept is health; for Fulford, it is illness. Beginning with either of these concepts is sometimes said to allow for a broader and richer conception of the health concepts than Boorse's theory allows, while avoiding the excesses of the WHO definition.[39]

Nordenfelt's Holistic Theory

In his monograph *On the Nature of Health* (originally published in 1987), Lennart Nordenfelt first set out what he calls a "holistic theory" of health, developed and defended in many subsequent publications.[40] In contrast to Boorse's understanding of disease and health in terms of statistical biological norms and deviations therefrom, Nordenfelt articulates what he takes to be an older conception of health in terms of *welfare*. This, he says, could emphasize either well-being or ability; of these two aspects, he gives the more central place to ability, leading him to define health thus:

> P is completely healthy, if and only if P has the ability, given standard circumstances, to realize all his or her vital goals.
>
> P is unhealthy (or ill) to some degree if and only if P, given standard circumstances, cannot realize all his vital goals or can only partly realize them.[41]

Nordenfelt makes an important distinction between "first- and second-order" ability:[42] first-order ability is the ability to do something, whereas second-order ability is the ability to *acquire* the ability to do something: for example, someone who moves to another country might lack the ability to speak its language, and thus be inhibited from pursuing vital goals such as earning a living; but he or she might still have the second-order ability to

---

38. Thomas Schramme, "Lennart Nordenfelt's Theory of Health: Introduction to the Theme," *Medicine, Health Care, and Philosophy* 10 (2007): 3-4 (p. 3).
39. E.g., Law and Widdows, "Conceptualizing Health."
40. Lennart Nordenfelt, *On the Nature of Health: An Action-Theoretic Approach*, 2nd ed. (Dordrecht: Kluwer Academic, 1995).
41. Lennart Nordenfelt, "Concepts of Health and Their Consequences for Health Care," *Theoretical Medicine* 14 (1993): 277-285 (p. 280).
42. Nordenfelt, *On the Nature of Health*, pp. 49-53.

learn the language. The ability relevant to health is second-order: "To be healthy is to have, at least, a second-order ability to perform a certain set of actions. To be ill is to have lost or, in general, to lack one or more of these second-order abilities."[43]

In contrast to other accounts of ability,[44] Nordenfelt specifies that the ability relevant to health is the ability to realize "*vital* goals," by which he means "the set of goals which are necessary and together sufficient for [the person's] minimal happiness."[45] Some vital goals are "basic," in the sense that they are "vital irrespective of the person's own intentions";[46] the basic vital goals Nordenfelt identifies all have to do with survival, since survival is a *sine qua non* for the realization of any person's minimal happiness in the long run.[47] Nordenfelt holds, however, that apart from such basic vital goals, most will be determined by each agent him or herself. His theory thus places a high premium on the autonomously chosen goods and goals of individuals in determining what counts as health.

The proviso about "standard circumstances" is illustrated by the example of a schoolboy (presumably with little or no prior training) who is able to fly an airplane given careful instruction and close supervision from a qualified pilot: given "extraordinary opportunities" he is able to fly the plane, but this does not mean that in "standard circumstances" he would have that ability.[48] This proviso introduces an element of cultural and social relativity into the definition, because those circumstances include the "environmental and cultural background":[49] Jozsef Kovacs in-

---

43. Nordenfelt, *On the Nature of Health*, p. 53.

44. Nordenfelt distinguishes his own account on this score from Fulford, *Moral Theory and Medical Practice*, and David Seedhouse, *Health: Foundations of Achievement* (Chichester: Wiley, 1986). See Lennart Nordenfelt, "The Logic of Health Concepts," in *Handbook of Bioethics*, ed. George Khushf (Dordrecht: Kluwer Academic, 2004), pp. 205-222 (p. 209).

45. Nordenfelt, *On the Nature of Health*, p. 90.

46. Nordenfelt, *On the Nature of Health*, p. 91.

47. This of course assumes — not surprisingly — that happiness is only this-worldly. A Christian theological tradition, committed as it is to the view that the lives of the martyrs are among the most fulfilled of human lives, will have to reject that assumption. However, a question then arises about the relationship between this-worldly and eschatological happiness. In chapters 3 and 4 I shall defend an account in which "health" refers to a limited aspect of this-worldly or "penultimate" human flourishing (to borrow Dietrich Bonhoeffer's term), but points forward to the "ultimate" or eschatological promise of perfect fulfillment.

48. Nordenfelt, *On the Nature of Health*, pp. 47-48.

49. Nordenfelt, *On the Nature of Health*, p. 64.

terprets it as "the typical environment, in which most of the people in a given society live."[50]

If health is so defined, how are we to understand disease and other forms of disorder? According to Nordenfelt, a disease is "a bodily or mental process which is such that it tends to reduce its bearer's health," where "health" is understood according to the definition quoted above.[51] It should be emphasized that a condition need not *actually* reduce a given individual's health in order to be counted as a disease, because it might be "latent" or "aborted at an early stage."[52]

A number of commentators find Nordenfelt's account to be a considerable improvement on Boorse's biostatistical model: it reflects the realities of health care more convincingly and avoids Boorse's reliance on a biologically-based notion of natural function, which (as we have seen) the latter's critics find problematic.[53]

Nonetheless, Nordenfelt has attracted some critical comment of his own. While some commentators welcome the fact that he avoids the narrow limits of Boorse's theory, Thomas Schramme has argued that Nordenfelt's concept of health is *too* broad, tending like the WHO definition toward the medicalization of many areas of life. He supports this claim with the thought experiment of an aspiring athlete who is unable to reach an ambitious target of achievement in her chosen field, despite intensive training: on Nordenfelt's view she lacks the ability to achieve one of her identified vital goals, and must be considered unhealthy, even though on a commonsense view she is "plainly" healthy.[54] Nordenfelt's response is that there are instances of ill-health that "have more to do with people's unrealistic or even dangerous goal-setting than with disabilities in the traditional sense of the word,"[55] and the proper response to this athlete's unhappiness

---

50. Jozsef Kovacs, "The Concept of Health and Disease," *Medicine, Health Care, and Philosophy* 1 (1998): 31-39 (p. 36). As we shall see, Kovacs finds this "standard circumstances" proviso problematic.
51. Nordenfelt, "Concepts of Health," p. 280.
52. Nordenfelt, "Concepts of Health," p. 280.
53. Law and Widdows, "Conceptualizing Health," p. 306. Khushf, "An Agenda for Future Debate," p. 20. As noted below, however, Khushf nonetheless holds that Nordenfelt shares key assumptions with conventional medical practice that are called into question by the changing organizational and intellectual context of contemporary medicine.
54. Thomas Schramme, "A Qualified Defense of a Naturalist Theory of Health," *Medicine, Health Care, and Philosophy* 10 (2007): 11-17.
55. Lennart Nordenfelt, "Establishing a Middle-Range Position in the Theory of Health: A Reply to My Critics," *Medicine, Health Care, and Philosophy* 10 (2007): 29-32 (p. 30).

would be to try to persuade her to relinquish such unrealistic goals. However, this does raise a more deep-seated issue about the subjectivist nature of Nordenfelt's conception of vital goals, which (as noted above) are understood as necessary conditions for the individual's minimal happiness, and which — apart from basic goals related to survival — are autonomously chosen by the individual. Nordenfelt himself has acknowledged that further work is needed on the tension between his "want-satisfaction" conception of happiness and an Aristotelian understanding of happiness as *eudaimonia*.[56] This tension between subjective and objective conceptions of human flourishing or happiness will re-surface more than once in what follows — for example, in relation to Law and Widdows' "capability" approach drawing on the work of Amartya Sen and Martha Nussbaum, and Christopher Megone's Aristotelian account of health — and is clearly a matter of some interest for the theological account to be developed later in the book.[57]

Another area of critical discussion relates to Nordenfelt's proviso about "standard circumstances." Jozsef Kovacs, interpreting "standard circumstances" to mean "those external factors with which practically every member of a society is confronted . . . [which] must include the norms, institutions, ideologies, etc., of a society," takes Nordenfelt's concept to require adjustment to the social context in which the individual finds herself. Accordingly, he doubts that it gives sufficient critical purchase against social and ideological contexts that are distorted, unreasonable, or unjust.[58] For example, the idea of health fostered by Nazi ideology was hospitable to notions of racial hygiene and eugenics that counted *(inter alia)* the lives of Jews, Roma, and people with various impairments as "lives unworthy to be lived."[59] Or again, in some societies in which slavery was seen as a

---

56. Lennart Nordenfelt, *Health, Science, and Ordinary Language* (Amsterdam: Rodopi, 2001), pp. 118-119, quoted by Law and Widdows, "Conceptualizing Health," p. 306.

57. Some of the disability theology authors discussed below in chapter 3 identify this view of individual autonomous choice with an anthropology characteristic of modern liberal society and the market economy. They criticize this anthropology for its exclusionary tendencies, particularly in respect of people with cognitive impairments; see especially chapters 2 and 3 in Thomas E. Reynolds, *Vulnerable Communion: A Theology of Disability and Hospitality* (Grand Rapids: Brazos, 2008), and Hans S. Reinders, *Receiving the Gift of Friendship: Profound Disability, Theological Anthropology, and Ethics* (Grand Rapids: Eerdmans, 2008).

58. Kovacs, "The concept of health and disease," p. 36.

59. For one part of this story, see LeRoy Walters, "Paul Braune Confronts the National Socialists' 'Euthanasia' Program," *Holocaust and Genocide Studies* 21.3 (2007): 454-487.

respectable social institution, the term "drapetomania" was coined to describe a supposed form of insanity which prompted slaves to run away from their masters.[60] According to Kovacs, Nordenfelt's theory seems to imply that those unable to achieve minimal happiness in such societies should be considered unhealthy, whereas it would be better to describe the *societies* that espoused these norms and ideologies as unhealthy. Kovacs therefore prefers the more overtly value-laden terminology of adaptation to "*reasonable* social norms" or circumstances, a usage that Nordenfelt too has adopted in some of his more recent publications.[61] However, Kovacs' stipulations of reasonable social norms and a healthy society lead him some way back towards the WHO definition, raising again the specter of a totalizing definition of health that tends to medicalize many areas of life. Moreover, as he himself acknowledges, the concept of reasonable social norms is even less well-defined than those of health and disease.[62]

A third set of questions relates to the fact-value distinction. Law and Widdows read Nordenfelt as accepting the distinction between the concepts of disease (objective and value-free) and illness (subjective and value-laden); it is simply that Nordenfelt reverses Boorse's conceptual hierarchy such that disease is defined in terms of illness or health, rather than the other way round. Diseases are states of the body or mind that tend to obstruct health or cause illness.[63] Assuming this reading is correct, Nordenfelt's concept of disease might be vulnerable to Scott DeVito's critique of Boorse, noted earlier: that value-judgments cannot be excluded from the conceptualization of disease itself. A related point is argued by George Khushf: in common with traditional conceptions of medicine, Nordenfelt still assumes a rigid fact-value distinction; but the changing institutional and intellectual context of medicine in countries like the United States now makes the problems of such a distinction increasingly clear.[64] As we shall see later in the chapter, other philosophical accounts too will

60. H. Tristram Engelhardt Jr., "The Disease of Masturbation: Values and the Concept of Disease," *Bulletin of the History of Medicine* 48 (1974): 234-248.

61. Kovacs, "The Concept of Health and Disease," p. 36 (emphasis added); Law and Widdows, "Conceptualizing Health," p. 306.

62. Kovacs, "The Concept of Health and Disease," p. 38.

63. Lennart Nordenfelt, "The Concepts of Health and Illness Revisited," *Medicine, Health Care, and Philosophy* 10 (2007): 5-10 (p. 8); Law and Widdows, "Conceptualizing Health," p. 307.

64. Khushf, "An Agenda for Future Debate," pp. 24-26. For Nordenfelt's reply to Khushf, see Nordenfelt, "Establishing a Middle-Range Position," pp. 30-32.

raise questions about the place of a fact-value distinction in conceptualizing health and disease.

Fulford

Like Nordenfelt, the psychiatrist and philosopher K. W. M. (Bill) Fulford reverses Boorse's conceptual priority of health concepts, though in Fulford's model it is illness rather than health that is conceptually prior. He argues that what he calls "biological theories," such as that of Boorse, and "social theories," such as Nordenfelt's, are "structurally interdependent" — in other words, "the force of either depends on conceptual elements suppressed in the one theory but emphasized in the other."[65] That is to say, he holds — as noted earlier — that Boorse's theory is covertly evaluative. Conversely, social theories of health, because they focus on the subjective experience of illness, are faced with a "demarcation problem" of differentiating between illness and other forms of subjectively negative experience. To solve the demarcation problem, Nordenfelt has developed the concept of vital goals, as discussed above. However, according to Fulford, he is still faced with the problem of differentiating vital from other goals, and this distinction relies on his class of "basic vital goals" — those that must be held by all humans as *sine qua non* for the pursuit of whatever other vital goals they might choose. Since the basic vital goals identified by Nordenfelt are all concerned with survival, Fulford argues, he requires the biological criteria that in Boorse's theory serve as the criteria of disease. Fulford emphasizes that it is not a criticism of either theory to draw attention to this structural interdependence: "both represent detailed and comprehensive accounts of the areas of the conceptual structure of medicine with which they are primarily concerned, disease and health respectively."[66] However, he argues that this structural interdependence does suggest the value of developing a "bridge theory" to "make the connections between them transparent": namely, a theory of illness in which

> the experience of illness would be marked out from other negatively evaluated experiences, not (as in biological theories) by reference to

---

65. K. W. M. Fulford, "Praxis Makes Perfect: Illness as a Bridge between Biological Concepts of Disease and Social Conceptions of Health," *Theoretical Medicine* 14 (1993): 305-320 (p. 305).

66. Fulford, "Praxis Makes Perfect," p. 307.

underlying disturbances of function, nor (as in social theories) by reference to a particular class of the goals of action, but as a particular kind of "failure of action."[67]

Fulford holds that because the problems with which the philosophy of health care is concerned are practical problems, the ultimate test of a theory of health must be its effect on the practice of health care. He argues that his "bridge theory" offers identifiable benefits to clinical practice, medical education, and research. In relation to clinical practice, he notes the tendency toward "territorial disputes" in which the more "scientific" disciplines such as high-tech hospital medicine enjoy a higher status than those disciplines perceived as less scientific, such as primary care and psychiatry — a situation that in his (and others') view has a distorting effect on practice. By reversing the order of priority in Boorse's theory, so that illness becomes the logically prior concept, Fulford suggests that the disparity in status between primary care and hospital medicine will be corrected, because the primary care worker should now be understood not as scientifically inferior to her hospital colleague, but as the professional at "the conceptually (as well as empirically) tricky sharp end of health care."[68] Likewise, the concept of mental illness appears more value-laden and less scientifically exact than physical illness, not because the scientific foundations of psychiatry are more poorly developed than those of internal medicine, but because the experience of mental illness is intrinsically more "conceptually tricky": "our evaluations of, say, anxiety . . . are more open and variable than of [physical] pain."[69] In relation to education, Fulford cites a then-innovative example of medical education in which ethics, law, and communication skills were integrated with one another and with each area of clinical training.[70] In terms of research, Fulford claims that his theory suggests a research agenda for mental health that (among other things) can address some persistent conceptual problems in the description and classification of mental illnesses, in part by directing more attention to the phenomenology of illness.[71]

67. Fulford, "Praxis Makes Perfect," p. 308.
68. Fulford, "Praxis Makes Perfect," pp. 309-311 (p. 311).
69. Fulford, "Praxis Makes Perfect," p. 311.
70. Fulford, "Praxis Makes Perfect," p. 313; Tony Hope, K. W. M. Fulford, and Anne Yates, *The Oxford Practice Skills Course: Ethics, Law, and Communication Skills in Health Care Education* (Oxford: Oxford University Press, 1996).
71. Fulford, "Praxis Makes Perfect," pp. 314-316.

Neither Boorse nor Nordenfelt is persuaded by Fulford's critical comments on their own theories, nor by his offer of a bridge theory. Boorse has offered extensive rebuttals of his many critics, including Fulford, in which he continues to maintain that he does not covertly import evaluation into his concept of disease, and it is possible to give an objective, value-free account of disease based on statistical norms of species-typical function.[72] Nordenfelt registers agreement with many of Fulford's practical conclusions and much of what Fulford says about the importance of subjective illness. However, he insists that his theory is "holistic" (incorporating both "biological" and "social" elements), not a "social theory" as Fulford suggests. He is therefore unpersuaded of the structural interdependence that Fulford identifies between his theory and Boorse's, and of the need for a bridge theory between the two.[73] However, the conceptual priority given by Fulford to the lived experience of illness is an important development taken up in other accounts, such as the more extensive phenomenological analysis of S. Kay Toombs discussed in the next section. In some of his more recent publications, Fulford has moved in a more explicitly teleological direction, in dialogue with the Aristotelian account of Christopher Megone, which will be a major focus of discussion later in the chapter.[74]

The Phenomenology of Illness: S. Kay Toombs

Fulford's emphasis on the lived experience of illness is developed more extensively and richly by S. Kay Toombs in her phenomenological essay *The Meaning of Illness*.[75] Quite apart from its importance in its own right, Toombs's analysis is of interest for the points of contact it offers with phenomenological perspectives on disability, to be surveyed in the next chapter, and with aspects of the theological discussion in later chapters. Toombs's starting point is her own experience of multiple sclerosis, of which she reports that "in discussing my illness with physicians, it has often seemed to me that we have been somehow talking at cross purposes,

---

72. Notably Boorse, "A Rebuttal on Health."

73. Lennart Nordenfelt, "On the Relation between Biological and Social Theories of Health: A Commentary on Fulford's 'Praxis Makes Perfect,'" *Theoretical Medicine* 14 (1993): 321-324.

74. Fulford, "What Is (Mental) Disease?"; Fulford, "Teleology without Tears."

75. S. Kay Toombs, *The Meaning of Illness: A Phenomenological Account of the Different Perspectives of Physician and Patient* (Dordrecht: Kluwer Academic, 1992).

discussing different things, never quite reaching one another."[76] A phenomenological perspective suggests that this experience of being at cross purposes arises because

> the physician and patient apprehend illness from within the context of separate worlds, each world providing its own horizon of meaning. Furthermore ... the experience of illness is such that it is particularly difficult to construct a shared world of meaning between physician and patient.[77]

This is the fundamental disjunction between physician and patient that she sets out to analyze by means of psychological phenomenology,[78] through detailed analyses of illness and the body drawing on Sartre, Merleau-Ponty, Husserl, and others.

In relation to illness, Toombs draws on the analysis of Jean-Paul Sartre, who identifies four levels at which the meaning of illness is constituted.[79] First, there is the level of what Toombs calls "pre-reflective sensory experiencing,"[80] such as the experience of pain. At this pre-reflective level, a pain in the eyes (for example) is not experienced as an object (pain) with a particular location; "[r]ather, pain *is* the eyes at this particular moment."[81] The location of pain *in,* say, the eyes or the stomach happens at the next level of meaning, when the pain is reflected upon as an object and identified with a "suffered illness."[82] At the third level, illness is conceived as "disease." This requires that the patient conceive of her body as an object, a "being-for-others," as Sartre puts it,[83] and to understand her illness as disease she will make use of knowledge acquired from others, such as ana-

---

76. Toombs, *The Meaning of Illness,* p. xi.
77. Toombs, *The Meaning of Illness,* p. 10.
78. Here she draws on the distinction in Husserl's work between "psychological" and "transcendental" phenomenology — the former making use of a "phenomenological reduction" or "bracketing" of the "outer world" so as to make explicit the individual's consciousness of that world, the latter performing the "phenomenological reduction" also on the individual consciousness in an attempt to gain access to the "ultimate structure of consciousness": Toombs, *The Meaning of Illness,* p. 121, n. 1.
79. Jean-Paul Sartre, *Being and Nothingness: An Essay on Phenomenological Ontology* (London: Methuen, 1957), pp. 303-359.
80. Toombs, *The Meaning of Illness,* p. 31.
81. Toombs, *The Meaning of Illness,* p. 31.
82. Sartre, *Being and Nothingness,* p. 356.
83. Sartre, *Being and Nothingness,* part 3 (title).

tomical and physiological knowledge about the structure and functioning of the eyes, stomach, and so forth. These first three levels may all be levels of the patient's apprehension of pain (though not every experience of pain will be apprehended as an illness or a disease). Sartre's fourth level, the "disease state," represents the physician's apprehension of the patient's condition, in which it is conceived in terms of biological cause and effect. It must be emphasized that for Sartre, the third and fourth levels ("disease" and "disease state" respectively) are distinct and different. "Disease" is one level at which the patient's *subjective* experience may be represented, and this experience is by definition not directly accessible to the physician; nor can the physician's knowledge of the "disease state" be directly experienced by the patient. A patient with multiple sclerosis, for example, "does not experience the disruption of the nerve pathways directly," whereas "the physician regards the fundamental entity as *being* the lesion in the central nervous system."[84] This phenomenological analysis helpfully illuminates the distinction between "illness" and "disease" made in various accounts; in particular, it fills out Fulford's account of the distinction and offers some support to his proposal that "illness" should be given conceptual priority in understanding the health concepts.[85]

Toombs observes that "illness engenders a fundamental change in the manner in which the body is experienced," including an alienation from the lived body brought about by the patient's own objectification of his body, and intensified by "the physician's construal of the body as a scientific object."[86] Accordingly, the next stage of her phenomenological analysis is concerned with the phenomenology of the body and the changes in the apprehension of the body wrought by the experience of illness. As with Sartre's analysis of pain and illness, the first level is the pre-reflective one at which "I do not explicitly thematize my body *as* a body . . . [a]s an embodied subject I *AM* my body."[87] The lived experience of the body at this pre-reflective level is characterized by an integration of its parts and activities that Toombs, following Zaner, calls "contextual organization."[88] It also exhibits a kind of "primary" knowing of the world through "sensory-

---

84. Toombs, *The Meaning of Illness*, p. 41, emphasis original.
85. Fulford, "Praxis Makes Perfect"; see above, pp. 16-18.
86. Toombs, *The Meaning of Illness*, p. 49.
87. Toombs, *The Meaning of Illness*, p. 57, emphasis original.
88. Toombs, *The Meaning of Illness*, pp. 55-56, citing Richard M. Zaner, *The Context of Self: A Phenomenological Inquiry Using Medicine as a Clue* (Athens, OH: Ohio University Press, 1981).

motor experience ... prior to any act of reflection or conceptualization,"[89] and a kind of practical relation to the world in which "objects are apprehended as manipulatable or utilizable by the body."[90] The next level corresponds to what Sartre calls "being-for-the-Other":[91] I come to apprehend my body as an object either "in the gaze of the Other" or in "limit situations" when, in Shaun Gallagher's words, "the organism loses or changes its rapport with the environment."[92] This "loss of rapport" can be caused by negative or positive experiences — for example, by pain, tiredness, sport, or sexual arousal — but according to Toombs it always entails a kind of alienation: "the objectification of the body at the reflective level involves a disruption of the unity of lived body."[93]

Illness, according to Toombs, "strikes at the fundamental features of embodiment," so that at the pre-reflective level it "manifests itself essentially as a disruption of lived body.... First and foremost illness represents dis-ability, the 'inability to' engage the world in habitual ways."[94] One's lived spatiality and temporality change, so that both "contextural organization" and familiar ways of "being-in-the-world" (such as posture, gestures, and so forth) are disrupted. At the next level, that of the body as an object, Toombs observes that "[i]llness represents a 'limit situation' in which the body is apprehended both as a material, physical entity and as a being-for-the-Other."[95] In illness our bodies are disclosed to us, not merely as biological organisms (as in other "limit situations"), but as *malfunctioning* organisms: entities with processes over which we have limited control; the body is "transformed into a new entity, the 'diseased body.'"[96] A further level is highlighted by considering the difference between the patient's body as apprehended by the patient and by the physician, recalling Sartre's distinction between "disease" and "disease state." To the physician, "the fundamental entity is the body-as-scientific-object";[97] furthermore, the

---

89. Toombs, *The Meaning of Illness*, pp. 54-55.
90. Toombs, *The Meaning of Illness*, pp. 53-54, citing Maurice Merleau-Ponty, *The Phenomenology of Perception*, trans. C. Smith (London: Routledge and Kegan Paul, 1962).
91. Sartre, *Being and Nothingness*, p. 339.
92. Toombs, *The Meaning of Illness*, pp. 58, 59, quoting Shaun Gallagher, "Lived Body and Environment," *Research in Phenomenology* 16 (1986): 139-170.
93. Toombs, *The Meaning of Illness*, p. 61.
94. Toombs, *The Meaning of Illness*, p. 62.
95. Toombs, *The Meaning of Illness*, p. 70.
96. Toombs, *The Meaning of Illness*, p. 76.
97. Toombs, *The Meaning of Illness*, pp. 80-81.

physician's gaze, assisted by techniques of clinical examination, laboratory analyses, X-rays, and other imaging techniques, goes "inside" the body, rendering the outward appearance (so to say) transparent.

All of this has important consequences for the clinical encounter. With the objectification of the body in the clinical setting, "the unity of the lived body disintegrates and the body is alienated from the self" — which can engender a profound sense of loss of control and autonomy, so that "patients often simply hand over their bodies to the physician for treatment."[98] However, Toombs also indicates more positive ways in which a "healing relationship" can be developed between physician and patient, so as to counteract this disjunction and alienation.[99] A first step is for physicians simply to recognize and attend to the typical characteristics of the lived experience of illness — not, to be sure, discarding the scientific understanding of the "disease state," but seeking a fuller understanding by performing "a temporary 'shift in consciousness' . . . to a lifeworld interpretation of the patient's disorder."[100] Secondly, it is possible for physicians to develop an empathy for the lived experience of their patients. This is possible because we all have everyday experience of being confronted by our bodies as objects with physical limitations and a certain intractability; but it can become more profound for health professionals who have lived through significant injury or disorder themselves.[101] Further, it is possible to tell "clinical narratives" that attend not only to the biomedical features of the disease but also to the patient's lived experience of illness.[102] All of these can contribute to a healing relationship that is not only focused on

---

98. Toombs, *The Meaning of Illness*, p. 85, citing Drew Leder, "Medicine and Paradigms of Embodiment," *Journal of Medicine and Philosophy* 9 (1984): 29-43.

99. These are developed particularly in her final chapter: *The Meaning of Illness*, pp. 89-119.

100. Toombs, *The Meaning of Illness*, p. 97.

101. An impressive example is the neurologist Oliver Sacks's book *A Leg to Stand On* (New York: Summit, 1984), in which he tells how he sustained — and recovered from — a serious leg fracture in a mountain walking accident. While much of his experience as a patient was profoundly disempowering, there were moments of hope and encouragement, and at the close of the narrative he remarks that the experience "taught me, changed me as nothing else could. Now I *knew*, for I had experienced myself. And now I could truly begin to understand my patients" (p. 202).

102. Again, Oliver Sacks is a striking example: see, e.g., *The Man Who Mistook His Wife for a Hat and Other Clinical Tales* (New York: Summit, 1985). In a recent book, *The Mind's Eye* (London: Picador, 2011), he movingly interweaves narratives of both his own and others' lived experiences of illness, limitation, and vulnerability.

curing diseases but also on the relief of suffering and the care of the patient even when a cure is not possible.[103]

Toombs relates powerfully the alienating effect of experiencing one's body as Other in illness, medical diagnosis, and treatment, and the implications she draws out for clinical practice are salutary and important. However, there are limitations to this kind of phenomenological perspective. One cautionary note is sounded by the sociologist Gareth Williams, who argues that such perspectives can focus too much on individual lived experience and divert attention from societal and structural causes of alienation.[104] Another comes from the theologian Gerald McKenny, commenting particularly on Drew Leder and Richard Zaner (on both of whom Toombs also draws) in the light of Michel Foucault's analysis of "biopower." McKenny argues that phenomenologists like Leder and Zaner too easily

> assume that dehumanization is the worst thing modern medicine does to us and that to confess one's lived world to health professionals and to bring it under the discipline of a now greatly expanded range of therapeutic techniques is therefore unambiguously liberating . . . they are unable to realize that a medicine that is no longer "dehumanizing" may in the name of liberation subject the body to relations of power that would then inscribe even the patient's lived world into the processes of normalization that serve the requisites of a certain kind of society.[105]

A Capability Approach: Law and Widdows

A contrasting (though quite possibly complementary) approach adopted by a few authors is to conceptualize health in terms of the "capability approach" associated with Amartya Sen, Martha Nussbaum, and others, developed originally in the context of development economics. Sen's version

---

103. See, further, Eric J. Cassell, *The Nature of Suffering and the Goals of Medicine*, 2nd ed. (New York: Oxford University Press, 2004).

104. Gareth Williams, "Theorizing Disability," in *Handbook of Disability Studies*, ed. Gary L. Albrecht, Katherine D. Seelman, and Michael Bury (Thousand Oaks, CA: Sage, 2001), pp. 123-144 (pp. 132-133).

105. Gerald P. McKenny, *To Relieve the Human Condition: Bioethics, Technology, and the Body* (Albany: State University of New York Press, 1997), p. 197.

of the approach is informed by a critical stance towards other accounts (for example, utilitarian and Rawlsian) of distributive justice and human well-being, which he argues use inappropriate metrics and criteria to assess development. Instead, he advocates an approach based on "some notion of 'basic capabilities': a person being able to do certain basic things."[106] The approach is built on the concept of "functionings" — things that a person manages to be or to do in the course of living his or her life. In a more recent paper Sen summarizes the approach as follows:

> The *capability* of a person reflects the alternative combinations of functionings the person can achieve, and from which he or she can choose one collection. The approach is based on a view of living as a combination of various "doings and beings," with quality of life to be assessed in terms of the capability to achieve valuable functionings.[107]

Some functionings, such as the ability to be well nourished and properly sheltered, will be "basic" and strongly valued by everyone; others, such as the ability to achieve social integration, are more complex — though, as Sen remarks, they might still be widely valued. Freedom is an important aspect of the capability approach: elsewhere he describes capability as "a set of vectors of functionings, reflecting the person's freedom to lead one type of life or another."[108]

In various places, Sen uses "being in good health" as an example of a basic functioning,[109] and following this lead, a few authors have adopted his approach to explore health and related concepts.[110] One interesting recent treatment is by Iain Law and Heather Widdows, who wish to go beyond viewing health as one of the functionings that contribute to capability, and to consider health itself "as 'capability' constructed from various

---

106. Amartya Sen, "Equality of What?" Tanner Lecture on Human Values, Stanford University, 22 May 1979. Available online at http://www.uv.es/~mperezs/intpoleco/Lecturcomp/Distribucion%20Crecimiento/Sen%20Equaliy%20of%20what.pdf.

107. Amartya Sen, "Capability and Well-Being," in *The Quality of Life*, ed. Martha C. Nussbaum and Amartya Sen (Oxford: Clarendon Press, 1993), pp. 30-53 (p. 31), emphasis in original.

108. Amartya Sen, *Inequality Reexamined* (Cambridge, MA: Harvard University Press, 1992), p. 40.

109. E.g., Sen, "Capability and Well-Being," p. 31.

110. E.g., M. A. Verkerk, J. J. V. Busschbach, and E. D. Karssing, "Health-Related Quality of Life Research and the Capability Approach of Amartya Sen," *Quality of Life Research* 10 (2001): 49-55.

possible functionings, in other words as a collection of valuable components which can be constructed in a variety of ways."[111]

Law and Widdows argue that this approach has several advantages over other theories of health and disease.[112] One is that it avoids both the narrowness of a theory such as Boorse's and the excessive broadness of the WHO definition. Another is that it steers a middle course between subjectivist theories such as Nordenfelt's and objectivist accounts such as Boorse's, avoiding (they claim) the problems of both and filling out Nordenfelt's somewhat indeterminate notion of "vital goals." In particular, because the health "capability set" is composed of a number of different functionings, the selection of those functionings can reflect the needs and values of individuals and their social contexts. It can thus take account of the contextual character of health, and this makes it serviceable for the assessment of health globally. Finally, Law and Widdows hold that it can capture the important insights in the disease-illness distinction made by many authors — and indeed the tripartite distinction of health, disease, and sickness[113] — without tying those insights to an overly rigid theoretical structure that might be somewhat detached from concrete experience.

This capability account offers a valuable perspective, and suggests informative connections with the discussion of disability in the next chapter and the theological account to be developed in later chapters.[114] One area that merits critical attention concerns what Sen calls the "incompleteness" of the capability approach,[115] by which he means that by design, it does not give one exclusive account of the human good or stipulate what functionings are

---

111. Law and Widdows, "Conceptualizing Health," p. 311. They acknowledge that there is more work to be done to articulate the relationship between health, understood as a capability, and other capabilities (p. 311 n. 3); this is one expression of a more general question (raised above, for example, in relation to the WHO definition, and later in relation to Megone's Aristotelian account) about the relationship between health and other human goods.

112. Law and Widdows, "Conceptualizing Health," pp. 311-312.

113. See Bjørn Hofmann, "On the Triad Disease, Illness, and Sickness," *Journal of Medicine and Philosophy* 27.6 (2002): 651-673.

114. In relation to disability, Law and Widdows offer some useful comments on the interaction of social context with various impairments: "Conceptualizing Health," p. 313. In relation to the theological account developed in later chapters, there are obvious connections with Karl Barth's description of health as "capability, vigor, and freedom . . . the integration of the organs for the exercise of psycho-physical functions"; *Church Dogmatics*, 13 vols., ed. and trans. Geoffrey W. Bromiley and Thomas F. Torrance (Edinburgh: T&T Clark, 1956-75), III.4, p. 356; hereafter *CD*. See further below, pp. 131-141.

115. Sen, "Capability and Well-Being," pp. 48-49.

needed for a flourishing human life. Law and Widdows argue that this gives their capability account of health an inherently pluralistic and flexible character, which makes it usable in a wide range of different contexts and in connection with different value-systems.[116] For some purposes (such as the assessment of global health inequalities) this has obvious advantages. In other ways, however, this "incompleteness" and pluralistic character could mean that it suffers from some of the same difficulties noted with other accounts that rely on socially-determined and contextually relative norms and values. A related point is made (in relation to Sen's capability approach generally) by Martha Nussbaum, who recognizes the compatibility of Sen's approach with an Aristotelian account of virtue and the good, but argues that

> Sen needs to be more radical ... in his criticism of the utilitarian accounts of well-being, by introducing an objective normative account of human functioning and by describing a procedure of objective evaluation by which functionings can be assessed for their contribution to the good human life.[117]

Sen's own response to that suggestion is that he has "no great objection to anyone going on that route," but that "the use of the capability approach as such does not require taking that route, and the deliberate incompleteness of the capability approach permits other routes to be taken which also have some plausibility."[118] In relation to conceptions of health, however, Nussbaum's question about the need for a normative account of human functioning and its contribution to the good life has been raised also by others, as we shall see.

The Harmful Dysfunction Model: Jerome C. Wakefield

Since the early 1990s, one of the more important and influential accounts has been Jerome Wakefield's model of *disorder as harmful dysfunction*. This

---

116. Law and Widdows, "Conceptualizing Health," pp. 311-314.

117. Martha Nussbaum, "Nature, Function, and Capability: Aristotle on Political Distribution," *Oxford Studies in Ancient Philosophy*, supp. vol. (New York: Oxford University Press, 1988), p. 176, quoted by Sen, "Capability and Well-Being," p. 47. See also Martha Nussbaum, "Non-Relative Virtues: An Aristotelian Approach," in Nussbaum and Sen, *The Quality of Life*, pp. 242-276.

118. Sen, "Capability and Well-Being," p. 47.

is a particularly interesting contribution to the debate for two reasons. First, it lays out more fully, explicitly, and rigorously the evolutionary foundations implicit in other biological accounts of disease such as Boorse's biostatistical theory.[119] It thus offers one of the best test-cases of the project to conceptualize health, disease, and disorder in terms of the biological underpinnings of modern medicine. Secondly, as will be seen in this and later sections, it brings into sharp focus the question of teleological explanations of natural functions, and by extension, the prospects for a teleological account of health concepts. It is therefore worth dwelling in some detail on Wakefield's model.[120]

Wakefield uses the term "disorder" rather than "disease" partly because his primary focus is psychiatry, where "disorder" is the more commonly used term, but also on the grounds that "disorder" is a more inclusive term, encompassing medical conditions (such as traumatic injuries) whose nature or etiology makes it implausible to describe them as "diseases."[121] In the 1992 paper in which he first proposes his model, he is animated by concerns shared with some of the authors already surveyed — for example, the conceptual difficulties involved in classifying mental disorders, the problem of distinguishing between disorders and non-disorders, and the dangers of abuse, such as the expression of social prejudices in diagnosis and the stigmatization of socially undesirable forms of behavior.[122] He finds existing accounts, including Boorse's biostatistical theory and Thomas Szasz's skeptical "antipsychiatry" stance that mental disorder is a myth, unsatisfactory for various reasons.[123] Instead he defines disorder as "harmful dysfunction," where

---

119. For another evolutionary account, see Mahesh Ananth, *In Defense of an Evolutionary Concept of Health: Nature, Norms, and Human Biology* (Aldershot: Ashgate, 2008).

120. Indeed, the length and exhaustive detail of many of his own papers makes it rather difficult *not* to.

121. Jerome C. Wakefield, "Aristotle as Sociobiologist: The 'Function of a Human Being' Argument, Black Box Essentialism, and the Concept of Mental Disorder," *Philosophy, Psychiatry, and Psychology* 7.1 (2000): 17-44 (pp. 19-20).

122. Jerome C. Wakefield, "The Concept of Mental Disorder: On the Boundary between Biological Facts and Social Values," *American Psychologist* 47.3 (1992): 373-388 (pp. 373-374).

123. Wakefield, "The Concept of Mental Disorder," pp. 374-380. Szasz's view of "mental disorder" as a myth is based on the claim that mental disorders, in contrast with physical disorders, cannot be identified with bodily lesions, but Wakefield argues that physical disorders cannot either. He rejects biostatistical models like Boorse's on the grounds that they equate disorders with negative effects on survival and fertility, an equation to which it is easy to find counter-examples. However, as I noted above in the discussion of Boorse, this kind of objection to his model might not be as telling as Wakefield thinks.

dysfunction is a scientific and factual term based in evolutionary biology that refers to the failure of an internal mechanism to perform a natural function for which it was designed, and *harmful* is a value term referring to the consequences that occur to the person because of the dysfunction and are deemed negative by sociocultural standards.[124]

Wakefield explains the design of internal mechanisms in neo-Darwinian evolutionary terms. A heritable characteristic will tend to become more common in a population over time if it increases (however slightly) the *inclusive fitness* of its bearers,[125] because the genetic factors that give rise to that characteristic will persist and spread in a population if individuals with those genetic factors on average produce more offspring than those without them.[126]

He makes two key points about the concept of a natural function.[127] First, functional explanation (whether of artifacts or of natural structures or mechanisms) is an unusual kind of causal explanation in that effects explain their causes. In the case of an artifact, the explanation lies in the prior

---

124. Wakefield, "The Concept of Mental Disorder," p. 374.

125. The concept of "inclusive fitness" is used to indicate that there is a variety of ways in which a heritable trait can increase in frequency in future generations. The most obvious is by increasing the reproductive success of individuals that bear the trait: that is, causing them to produce more offspring than others that lack it. But another way is by causing some of its bearers to support the reproductive success of *other* individuals that share the same trait. The classic example is the puzzle of how "altruistic" behavior (in the sense of behavior that reduces the individual's chances of survival and/or reproduction) can persist and spread in a population, even in species such as social insects where behavior is clearly not the product of conscious deliberation. William Hamilton's theory of "kin selection" and Robert Trivers's theory of "reciprocal altruism" are famous theoretical accounts of how altruistic behavior toward kin and non-kin, respectively, can increase the frequency of a "gene for" such behavior in the population. See, further, Neil Messer, *Selfish Genes and Christian Ethics: Theological and Ethical Reflections on Evolutionary Biology* (London: SCM, 2007), pp. 65-74, and references therein.

126. One complicating factor, which Wakefield acknowledges but disregards for the purposes of his model, is that natural selection is theorized by evolutionary biologists as operating at different levels: the gene, the organism, and (more controversially) the group. Wakefield chooses to discuss natural function in terms of selection operating at the level of the organism, on the grounds that a conceptual analysis of (mental) disorder is interested in dysfunctions operating at that level: "Aristotle as Sociobiologist," p. 43 n. 12. This is a defensible choice, though different views of the "levels of selection" debate will result in differences of detail in the way that evolutionary explanations of natural function and dysfunction are conceptualized. See, further, Messer, *Selfish Genes*, pp. 10-12, and references therein.

127. Wakefield, "The Concept of Mental Disorder," pp. 382-83.

purpose of its designer to make something that will produce the effect sought. In the case of natural functions, the mechanism's effect on the reproductive success of members of previous generations is the causal explanation for the presence of that same mechanism in the present generation. In earlier presentations of his model, he tends to regard this form of evolutionary causal explanation as a better alternative to both Aristotelian teleology and theological explanation:

> Evolutionary theory provides the only plausible scientific account that presently exists of how the natural functions of a mechanism can explain the existence and structure of a mechanism (earlier accounts of the same phenomenon included Aristotle's "final causes" and Aquinas's "God's intentions").[128]

More recently he has argued that his evolutionary account of natural function should be understood as a reading *of* Aristotelian teleology, a reading that he finds more persuasive than other interpretations of Aristotle.[129] This raises an important and interesting issue, to be revisited later, about the relationship between evolutionary and teleological accounts of health and human functioning.

Wakefield's second point about functional explanations is that "they can be plausible and very useful even when very little is known about the actual nature of a mechanism."[130] Part of the importance of this is that it saves the model from the (impossibly rigorous) requirement to provide an *actual* evolutionary explanation for the existence of a structure or mechanism if its failure is to be identified as a dysfunction. Rather, if a mecha-

---

128. Jerome C. Wakefield, "Evolutionary versus Prototype Analyses of the Concept of Disorder," *Journal of Abnormal Psychology* 108.3 (1999): 374-399 (p. 375).

129. Wakefield, "Aristotle as Sociobiologist." He does not, however, seem to have relinquished the assumption that evolutionary biology and theology are rival explanatory accounts of design in nature, and that theology has been superseded by evolution: cf. p. 43, n. 13. This is a common misconception among writers on evolution, but a misconception nonetheless, arising from the assumption that the only explanatory questions to be asked about natural phenomena are questions about efficient causation, an assumption that will be questioned below. See further Messer, *Selfish Genes*, pp. 49-53. This point will of course become important later in the book in the development of a theological account of health concepts, when I shall insist that a theological account of the kind I aim to articulate is perfectly able to assimilate explanatory insights from evolutionary biology, along with many other fields of human experience and understanding.

130. Wakefield, "The Concept of Mental Disorder," p. 382.

nism provides a clear benefit that depends on a complex and precise set of interactions, it is reasonable to infer that the benefit is not accidental, and — in the absence of other plausible explanations — that the benefit is a natural function of the mechanism.

> For example, it cannot be merely a happy accident that the eyes enable us to see. . . . The eyes therefore must exist in part because they enable us to see; that is, the fact that the eyes provide sight must somehow enter into the explanation of why we have eyes. This makes seeing a function of the eyes. Obviously, one can go wrong in such explanatory attempts; what seems nonaccidental may turn out to be accidental. But often one is right, and functional explanatory hypotheses communicate complex knowledge that may not be so easily and efficiently communicated in any other way.[131]

In a more recent development of the model, he has extended this point to the claim that *natural function* is what he calls a "black box essentialist" scientific concept.[132] This requires a little explanation. According to Wakefield, when we consider a particular kind of being, there will be a "base set" of members, which we initially identify as members of that kind because they share certain observed properties. For example, we might initially identify water operationally as a clear, colorless, thirst-quenching liquid, and different samples will be identified as liquid water if they possess the same observed properties. However, possession of these properties is not the *criterion* by which we classify other things as members of the same kind. The criterion is that they share the same essence as the members of the base set, even if some of their properties are different. For example, neither ice nor steam is a clear, thirst-quenching liquid, but both count as water. The reason is that they share an underlying essence with liquid water, and that essence turns out to be the molecular structure H_2O.

By analogy, the criterion for counting something as a natural function is that it shares a "hypothesized underlying property" with members of the "base set" — those features of an organism that are unproblematically recognized as natural functions. In other words, "a natural function is a nonaccidental effect that possesses the same essential property as the benefits

---

131. Wakefield, "The Concept of Mental Disorder," p. 383.
132. Wakefield, "Aristotle as Sociobiologist," pp. 35-39.

provided by the eyes, hands and so on."[133] That essential property turns out to be that such effects are the products of evolutionary design.

Like a number of the other accounts surveyed in this chapter, Wakefield's raises the issue of the fact-value distinction, and more generally of the relationship between factual and evaluative components in an account of health. Wakefield agrees with other critics that Christopher Boorse's biostatistical theory is unsuccessful in excluding value from the concept of disease, and his own model of disorder is an attempt to take seriously the insight that "[t]o have a disease, an illness or a disorder is necessarily to have a *(prima facie)* negative condition."[134] So his concept of disorder combines factual and evaluative components; but he still maintains a sharp fact-value distinction, maintaining that the "dysfunction" component of the definition is scientific and strictly value-free, while the evaluative element is located in the criterion of "harm."[135]

This raises two critical questions. One is whether Wakefield is correct that the concept of "dysfunction" is value-free. Scott DeVito, in an argument parallel to one of his criticisms of Boorse, holds that it is not: the choice of an evolutionary criterion to define "dysfunction" is itself value-laden,

> because it [relies on the premise] that the only good explanation of a body-state functioning in a particular manner is an evolutionary explanation. But that need not be true. We can provide analyses of bodystates on the biochemical level, at the social level . . . at the anatomical level, and so on. Our decision to use any of these types of definitions of functionality will be based, in part, upon what about the entity under study is of interest to the investigator. . . . And, as such, the choice is value-laden.[136]

---

133. Wakefield, "Aristotle as Sociobiologist," p. 36.

134. Wakefield, "Aristotle as Sociobiologist," p. 19. The qualification *prima facie* is routinely introduced into value-laden theories of disease, illness or disorder, because the evaluative criterion is only that these conditions are *usually* negatively evaluated. There could be circumstances in which a disorder was not a bad thing to have: cowpox is a disease, but in times and places where smallpox was widespread and there was no artificial vaccine, cowpox could be a very *good* condition to contract.

135. This distinction is defended at length, for example, against Christopher Megone in "Aristotle as Sociobiologist."

136. DeVito, "On the Value-Neutrality of the Concepts of Health and Disease," p. 554.

However, this is less persuasive than DeVito's parallel critique of Boorse for the simple reason that all of the levels of explanation to which he alludes (at any rate, all of the levels of *biological* explanation — depending on exactly what is meant, "social" is a less clear-cut case) do in some way presuppose an evolutionary paradigm. It might not always be in the foreground of detailed research in, say, biochemistry, but it nonetheless supplies the conceptual framework within which almost any biological research program can be located. The famous assertion of the geneticist Theodosius Dobzhansky that "nothing in biology makes sense except in the light of evolution" would command widespread assent among working biologists.[137] So the choice of an evolutionary criterion for identifying natural functions and dysfunctions does not seem to be value-laden in the sense that DeVito claims, namely that it depends on the interests of particular scientific investigators.

A different critique of Wakefield's claim that "natural function" and "dysfunction" are value-free concepts comes from Christopher Megone, who argues in Aristotelian terms that "natural function" must be understood as a teleological concept, and this is inescapably evaluative.[138] The argument between Megone and Wakefield as to whether Aristotelian teleology can be understood in evolutionary and/or value-free terms is of central interest for the present project, and will be taken up again in a later section.

The second critical issue raised by the separation of fact and value within Wakefield's definition of a disorder concerns the nature of the evaluative component, the criterion of *harm*. He has less to say about the "harm" criterion than about the evolutionary "dysfunction" component of his definition, but as we have seen, he understands "harmful" to be "a value term based on social norms."[139] In common with others such as Nordenfelt, Wakefield therefore holds the evaluative element to be socioculturally relative, though it is less clear whether he also agrees with Nordenfelt that this element will be subjective in the sense of being based on individuals' autonomously chosen goals. In the discussion of Nordenfelt above, the concerns of critics such as Kovacs about the cultural relativity of this evaluative component have already been noted; the issue

---

137. Theodosius Dobzhansky, "Nothing in Biology Makes Sense except in the Light of Evolution," *American Biology Teacher* 35 (1973): 125-129; reprinted in Mark Ridley, ed., *Evolution*, Oxford Readers, 2nd ed. (Oxford: Oxford University Press, 2004), pp. 400-409.

138. Megone, "Aristotle's Function Argument"; "Mental Illness, Human Function, and Values."

139. Wakefield, "The Concept of Mental Disorder," p. 373.

of whether the values embodied in health concepts are objective or subjective, and universal or particular to social context, will recur at intervals throughout the present chapter and will clearly be of importance to the development of a theological account later in the book.

Another important area of critical discussion has to do with the way in which Wakefield's concepts of "natural function" and "dysfunction" draw on evolutionary theory, and whether they depend on a particular interpretation of evolution. These issues are raised, for example, by Scott Lilienfeld and Lori Marino, who criticize Wakefield's use of evolutionary theory in three ways. First, they argue that it suffers from the shortcomings of an "adaptationist" view of evolution, which assumes that all features of an organism are adaptations shaped directly by evolutionary selection pressures. "In contrast to Wakefield's assertions," they remark, "many important physical and mental systems were probably not designed directly by evolution to perform a given function."[140] They support this claim by reference to various evolutionary theorists, notably Stephen Jay Gould and his concept of "exaptations" (features that were not directly selected for, but arose as byproducts of other changes that were the direct outcomes of natural selection).[141] Gould identifies two kinds of exaptation: "secondary adaptations," which became adaptive for their bearers at some time after they originally appeared, and "adaptively neutral exaptations," which did not. Lilienfeld and Marino suggest that language is a probable example of the first kind of exaptation, a secondary adaptation, and they allow that such features can be accommodated within Wakefield's concepts of function and dysfunction. However, the second kind (adaptively neutral exaptations) pose a more serious challenge. Lilienfeld and Marino propose a number of examples, including religion, arithmetical ability, music and art. While some at least of their examples are questionable,[142] the general point that

---

140. Scott O. Lilienfeld and Lori Marino, "Mental Disorder as a Roschian Concept: A Critique of Wakefield's 'Harmful Dysfunction' Analysis," *Journal of Abnormal Psychology* 104 (1995): 411-420 (p. 412).

141. Stephen Jay Gould, "Exaptation: A Crucial Tool for Evolutionary Analysis," *Journal of Social Issues* 47 (1991): 43-65.

142. In general, theories of the evolutionary origins of mental and behavioral capacities are often highly contested and (in the nature of the case) drastically under-determined by hard evidence. For example, contra Lilienfeld and Marino, there are evolutionary theories of the origin of religion that would regard it as either a secondary or a direct adaptation, such as David Sloan Wilson, *Darwin's Cathedral: Evolution, Religion, and the Nature of Society* (Chicago: University of Chicago Press, 2003). For a philosopher of religion's critical survey

there may well be important human abilities that are neither direct nor secondary adaptations seems plausible, and they argue that extending Wakefield's concept of natural function to include such cases would have odd consequences. For example, allowing for the sake of argument that "religion" *was* a non-adaptive exaptation, if such features were included in an extended form of Wakefield's definition, then this would seem to make atheism a dysfunction and — in some social contexts — a disorder.[143]

Second, many mental disorders seem to represent extremes on a range of functioning, rather than all-or-none conditions, so that Wakefield understands some dysfunctions as "graded deviation from an evolutionarily designed level of natural functioning."[144] The problem with this, according to Lilienfeld and Marino, is that "evolution rarely designs universal set points across individuals for the functioning of biological systems; there is almost always considerable variability around the mean of the population distribution of a system's responses."[145] So it might be difficult or impossible to define an "evolutionarily designed level of natural functioning," in which case an evolutionary concept of function might not supply clear criteria for distinguishing disorders from non-disorders.

Third, many disorders, so far from being *failures* of evolutionarily designed natural functions, seem to be *adaptive* responses of the organism's systems to certain threats or insults. For example, symptoms of viral infection such as fever, cough, and sneezing, and symptoms of food poisoning such as diarrhea and vomiting, appear to be adaptive responses designed to expel infectious organisms or toxins from the body; in some cases, infections persist for longer if their symptoms are suppressed by medical means. This claim is most clear-cut in such examples of physical illness, but Lilienfield and Marino propose that certain mental disorders, such as some anxiety disorders and phobias, are likewise adaptive responses to traumatic or threatening situations.[146]

Wakefield has published an extensive response to Lilienfeld and Marino's critique.[147] In brief, in response to their objection about exaptations,

---

of some evolutionary and cognitive-science theories of religion, see Gregory R. Peterson, "Are Evolutionary/Cognitive Theories of Religion Relevant for Philosophy of Religion?" *Zygon* 45.3 (2010): 545-557.

143. Lilienfeld and Marino, "Mental Disorder as a Roschian Concept," p. 413.
144. Lilienfeld and Marino, "Mental Disorder as a Roschian Concept," p. 414.
145. Lilienfeld and Marino, "Mental Disorder as a Roschian Concept," p. 414.
146. Lilienfeld and Marino, "Mental Disorder as a Roschian Concept," pp. 415-416.
147. Wakefield, "Evolutionary versus Prototype Analyses."

he argues that almost all biological exaptations involve *some* modification of the exapted structures under the influence of natural selection to perform their new function, so evolutionary design is required to explain the present function. For example, it is thought that birds' feathers were originally a form of thermal insulation, later exapted for the purpose of flight; but almost certainly some structural alterations would have been required (and selected for) to make feathers suitable for this new purpose.[148] By contrast, Wakefield allows that many *cultural* exaptations, such as reading, are genuinely non-adaptive: they involve the purposeful use of an ability for a purpose other than the one for which it was selected. However, he denies that this undermines his "harmful dysfunction" analysis, because "[t]here is no requirement that the harm must be identical to the failed natural function."[149] A reading disorder, if caused by a failure of some mental mechanism, does harm in some contexts for social and cultural, not biological, reasons, and therefore in those contexts is a genuine disorder. Against Lilienfeld and Marino's second point, that evolutionarily designed functions often span a range and therefore cannot supply clear criteria for distinguishing disorders from non-disorders, Wakefield acknowledges that the boundaries between disorder and normal functioning are often unclear, but argues that it is not the purpose of his model to supply sharply demarcated boundaries between function and dysfunction. Rather, his analysis is intended to "[explain] shared judgments about a range of important cases that clearly fall on one side or the other of the boundary."[150] To the third criticism, that some disorders are adaptive responses to threats, Wakefield replies that far from being "designed disorders," symptoms such as sneezing, coughing, and vomiting are adaptive (if unpleasant) *responses* to genuine disorders (such as respiratory tract or gastrointestinal infections) that disrupt adaptive biological mechanisms.[151]

It seems, on the whole, that Wakefield's model is able to resist the quite detailed criticisms leveled by Lilienfeld and Marino at his use of evolutionary theory, including the anti-adaptationist challenge drawn from evolutionary theorists such as Gould. There is potentially a more general difficulty with his evolutionary understanding of natural function in relation to mental disorder (though it would apply much less, if at all, to bodily dis-

---

148. Wakefield, "Evolutionary versus Prototype Analyses," pp. 380-381.
149. Wakefield, "Evolutionary versus Prototype Analyses," p. 382.
150. Wakefield, "Evolutionary versus Prototype Analyses," p. 379.
151. Wakefield, "Evolutionary versus Prototype Analyses," pp. 391-392.

eases). That is the argument from some biologists that evolutionary selection pressures have not in general shaped specific mental functions or adaptations; what evolution has produced is a brain with extraordinary flexibility and plasticity, whose specific mental functions are more directly attributable to its development under sociocultural influences than to biologically inherited predispositions shaped by natural selection.[152] Even an extreme version of this argument might not prove fatal to the harmful dysfunction model, but it might call into question how many harmful mental conditions could properly be called mental disorders attributable to failures of evolutionarily designed functions.

The Prototype Model: Lilienfeld and Marino

In the previous section, Scott O. Lilienfeld and Lori Marino's critique of the evolutionary basis of Wakefield's harmful dysfunction model was surveyed in some detail.[153] Their detailed critique of Wakefield leads them to conclude that "criterial" models such as his fail to establish a correspondence between the conceptions of disorder proposed in the models and some entity in nature. They argue that given the difficulty of establishing such a correspondence, there are grounds for thinking that it does not exist: there is nothing in the natural world that corresponds to the concept of "disorder." (They do not mean to say that particular entities classified as disorders, such as schizophrenia, do not exist in nature; simply that the classification "disorder" does not correspond to a natural kind.) Accordingly, they propose that "disorder" is a concept of the sort described by the psychologist Eleanor Rosch:[154] "mental constructions that are typically used to categorize entities in the natural environment (e.g., bird, fruit, mountain)."[155] Roschian concepts are organized around prototypical examples that display all of the constitutive features of the concept; members of the category are identified by their family resemblance (so to speak) to the prototype, but some will resemble it more closely than others. This means that "fuzzy" boundaries are a characteristic feature of Roschian concepts. Furthermore, since Roschian concepts are mental constructs, different individuals' ver-

---

152. See, e.g., Steven Rose, *The Future of the Brain* (Oxford: Oxford University Press, 2005), pp. 94-105.
153. Lilienfeld and Marino, "Mental Disorder as a Roschian Concept."
154. E.g., Eleanor R. Rosch, "Natural Categories," *Cognitive Psychology* 4 (1973): 328-350.
155. Lilienfeld and Marino, "Mental Disorder as a Roschian Concept," p. 416.

sions of the concept of disorder "almost certainly exhibit less than perfect overlap."[156] Because of these two features, Lilienfeld and Marino suggest, it is to be expected that some pyschological conditions such as personality disorders will lie on the (inherently unclear) boundaries of the concept, and that there will be intractable disagreements about which conditions should be included in the category of "disorder."

Their Roschian understanding of the concept of mental disorder leads Lilienfeld and Marino to think that "Wakefield's conceptualization of disorder [as harmful dysfunction] may actually prolong scientific debate on a fundamentally non-scientific issue."[157] An example is the time and energy spent on deciding which conditions are to be classified as disorders in successive revisions of psychiatric diagnostic manuals such as the American Psychiatric Association's *Diagnostic and Statistical Manual (DSM)*.[158] They argue for a different kind of *DSM* revision process in which the aim is not to classify conditions as disorders or non-disorders, but instead to produce "a careful compilation of well-validated conditions that are currently deemed to require intervention . . . from mental health professionals," based in part on each condition's "harmfulness" to the individual, society, or both.[159] To the objection (familiar from anti-psychiatry critiques such as that of Szasz) that this opens the diagnosis of mental disorder up to social and political manipulation directed at controlling behavior deemed problematic, they offer the — perhaps over-sanguine — response that such social and political influence has often been a feature of such judgments, and it would be better made explicit and exposed to proper scrutiny and debate.

In essence, then, Lilienfeld and Marino's Roschian model of the concept of disorder is a sophisticated defense of what Wakefield calls a "pure value" concept of mental disorder, of a kind earlier proposed by H. Tristram Engelhardt and Peter Sedgwick, among others.[160] To repeat, this does

---

156. Lilienfeld and Marino, "Mental Disorder as a Roschian Concept," p. 417.

157. Lilienfeld and Marino, "Mental Disorder as a Roschian Concept," p. 418.

158. The current version is American Psychiatric Association, *Diagnostic and Statistical Manual of Mental Disorders, Fifth Edition (DSM-5)* (Arlington, VA: American Psychiatric Publishing, 2013). For information about the revision process leading to the publication of *DSM-5*, see American Psychiatric Association, *DSM-5 Development*, online at http://www.dsm5.org/Pages/Default.aspx.

159. Lilienfeld and Marino, "Mental Disorder as a Roschian Concept," p. 418.

160. Wakefield, "The Concept of Mental Disorder," pp. 375-377; Engelhardt, "The Disease of Masturbation," pp. 234-248; Peter Sedgwick, *Psycho Politics* (New York: Harper and Row, 1982).

not entail denying that the entities classified as disorders (such as schizophrenia) are real entities: simply that their classification *as* disorders is a matter of pure value-judgment by individuals or societies. Part of Lilienfeld and Marino's case for such a view is that no "criterial" accounts of disorder had yet (at the time of writing their paper) succeeded. It was argued in the previous section that they have not succeeded in showing that Wakefield's evolutionary account of dysfunction (on which they train most of the critical firepower that they direct against Wakefield) fails, which somewhat weakens their own case.[161] However, the Roschian model does express two important insights: that categories such as "disease" and "disorder" seem to have irreducibly fuzzy boundaries, and that we often identify members of those categories at least in part by their similarity to what Lilienfeld and Marino call "prototypical" examples. It is possible, however, to retain both of these insights without being committed to Lilienfeld and Marino's conclusion that "the concept of disorder . . . lacks any clear-cut natural counterpart."[162] Wakefield himself argues that his evolutionary model is quite compatible with there being imprecise boundaries to the categories of dysfunction and disorder;[163] and in response to one of his recent critics, Lennart Nordenfelt remarks both that the concept of health has fuzzy boundaries and that it has "a conceptual torso that is given in ordinary language."[164]

A Teleological Account: Christopher Megone

Several of the accounts already surveyed, notably those of Boorse and Wakefield, rely on the concept of "natural function," defined in various

---

161. In addition to defending his own evolutionary model against Lilienfeld and Marino's specific criticisms, Wakefield argues at considerable length, by means of a series of "conceptual experiments," that the Roschian model has *less* explanatory power than his own: "Evolutionary versus Prototype Analyses," pp. 382-397. A conclusive judgment of this dispute would require an evaluation of each of these, which is beyond the scope of this chapter. In general terms, my view is that while some of these "experiments" are less persuasive than others, the most important critical questions to be put to Wakefield's account are concerned not with his evolutionary concept of dysfunction (which seems reasonably well-founded in terms of its own scientific paradigm), but with the relationship between fact and value in his account. These questions will be explored further in the next section, on teleology.

162. Lilienfeld and Marino, "Mental Disorder as a Roschian Concept," p. 416.
163. Wakefield, "Evolutionary versus Prototype Analyses," pp. 378-380.
164. Nordenfelt, "Establishing a Middle-Range Position," p. 31.

kinds of biological terms. Other accounts have a central focus on aims and goals — Nordenfelt's definition of health in terms of individuals' ability to realize their vital goals being a case in point. All of this raises a set of questions about the relationship of functions to goals, and the relationship of fact and value in the identification of each. One answer to these questions, proposed by Christopher Megone with specific reference to mental illness, takes the form of an Aristotelian teleological account of the natural function of a human being.[165] In Megone's reading of Aristotle, "facts" and "values" (to put the matter in modern, rather un-Aristotelian terms) are not sharply separated, as they are in most of the accounts surveyed thus far; and contra Boorse and Wakefield, Megone holds that the concept of a natural function is inherently "value"-laden.

Megone's starting point is Aristotle's attempt in the *Nicomachean Ethics* to clarify what is meant by *eudaimonia* ("happiness" or "flourishing"), which he does by asking about the function of a human being:

> For just as for a flute-player, a sculptor, or an artist, and, in general, for all things that have a function or activity, the good and the "well" is thought to reside in the function, so would it seem to be for man, if he has a function. . . . Or as eye, hand, foot, and in general each of the parts evidently has a function, may one lay it down that man similarly has a function apart from all these? What then can this be? Life seems to be common even to plants, but we are seeking what is peculiar to man. Let us exclude, therefore, the life of nutrition and growth. Next there would be a life of perception, but it also seems to be common even to the horse, the ox, and every animal. There remains, then, an active life of the element that has a rational principle.[166]

Megone offers a reading of this passage, with reference to Aristotle's discussion of natural kinds in the *Physics*,[167] in which a member of a natural kind (such as an oak tree) might undergo all kinds of changes. Some of these changes can be explained teleologically: they can be understood as part of a cycle of changes that contribute to the survival and persistence of the species. These are the changes that a good member of the species will

---

165. Megone, "Aristotle's Function Argument."
166. Aristotle, *Nicomachean Ethics*, trans. W. D. Ross, I.7, 1097b25-1098a4. Available online at http://classics.mit.edu/Aristotle/nicomachean.1.i.html.
167. Aristotle, *Physics*, trans. R. P Hardie and R. K. Gaye, II.8, 198b10-199b31. Available online at http://classics.mit.edu/Aristotle/physics.2.ii.html.

undergo — so a good acorn will grow into a vigorous sapling, then into a full-grown tree, which in turn will produce more acorns, and so forth.

According to Megone, if the human species is understood as a natural kind, which undergoes certain changes that are open to teleological explanation (as contributions to the survival and persistence of this particular natural kind), that helps to explain the argument about human functions quoted above. It is only possible to understand some of the changes that happen to parts of a human being (such as the eye, hand, or foot) in *functional* or teleological terms if they contribute to the function of the whole being. But in order to identify which of all the changes that happen to a human being are open to teleological explanation, Aristotle thinks it is necessary to identify what is distinctive to this particular (human) kind ("... we are seeking what is peculiar to man"). He therefore rules out properties that humans share with plants and animals, and is left with a *rational* life as the distinctive feature of the human species: "a good member of the human species will go through a cycle of changes which contribute to the developing and exhibiting of his potential for rationality. It is therefore changes that contribute to this cycle that are open to functional explanation."[168] It should be emphasized that Aristotle understands the life of reason very broadly: it includes both theoretical and practical rationality (and therefore the intellectual and moral virtues respectively), and will encompass a wide range of capacities that make human interpersonal life and sociality possible, such as empathy and language.[169]

In this perspective, illness and disease are to be understood as *failures of function:* "[failures] of certain changes whose occurrence would be teleologically explained as contributing to the cycle of changes exhibited in a good human life."[170] This, incidentally, helps to distinguish the functions of body parts from among the various things that they do. The reason that pumping blood, for example, can be recognized as the heart's function, whereas making a noise like a heart is not, is that pumping blood is recog-

---

168. Megone, "Aristotle's Function Argument," p. 195.
169. So Megone, "Mental Illness, Human Function, and Values," p. 53, in response to Wakefield, who (mistakenly, in my view) regards these as *other* candidates for teleological explanation, distinct from the life of reason: "Aristotle as Sociobiologist," p. 24.
170. Megone, "Aristotle's Function Argument," p. 195. He remarks that the same account applies to the concepts of disease and illness, and that both concepts must be understood as evaluative in the same way; though he does not really make clear whether the two concepts are distinct in any respect, or whether this account takes "illness" and "disease" to be synonyms.

nizable as a contribution to survival, and therefore a prerequisite for the living of a good human life.[171] It also shows that health, illness, and disease are inescapably evaluative concepts, since the functions of organs are recognizable as functions by virtue of their contribution to the cycle of changes characteristic of a *good* human life. Megone does, however, emphasize — in common with most other accounts, apart from the WHO definition — that health is only a part of human well-being or the good. Finally, Megone's account implies that physical and mental illnesses can be conceptualized in essentially the same way — as failures of function that incapacitate the living of a good human life, which is a *rational* life — though there will be differences in the ways in which physical and mental illnesses incapacitate the living of a rational life, and there may well therefore be differences in the kinds of treatment appropriate to each.[172]

Jerome Wakefield has extensively critiqued Megone's reading of Aristotle's function argument.[173] One aspect of his critique is that Megone's reading of Aristotle does not show why reason is the function of human beings. Wakefield offers an alternative reading, shaped by the presuppositions of his own evolutionary account of natural function, in which Aristotle becomes a kind of proto-sociobiologist.[174] Reason, on this account, serves as a "master function," regulating the expression of other human functions to enable the human being to live as a member of a society. Implicit in this analysis is a characteristic assumption of sociobiology, that humans have evolved to live in complex social structures — that is, their inclusive reproductive fitness is enhanced by living in such structures — and therefore a wide range of human traits and capacities, such as language, the expression of the emotions, and aspects of morality such as fairness and (reciprocal) altruism, can be explained as adaptations to enable this kind of social life. As already noted, Wakefield has a narrower under-

---

171. Megone, "Mental Illness, Human Function, and Values," p. 57. The example of the heart pumping blood and making heartbeat noises is attributed by Jerome Wakefield to Carl Hempel: Wakefield, "Aristotle as Sociobiologist," p. 21, citing Carl Hempel, "The Logic of Functional Analysis," in *Aspects of Scientific Explanation* (New York: Free Press, 1965).

172. Megone, "Aristotle's Function Argument," pp. 196-200.

173. Wakefield, "Aristotle as Sociobiologist." One aspect of this critique, which cannot detain us here, is that Megone fails to show why Aristotle might be justified in thinking that a human being *qua* human being has a function. Megone responds that this objection is based on a misreading of his, Megone's, account; see Wakefield, "Aristotle as Sociobiologist," pp. 22-24, and Megone, "Mental Illness, Human Function, and Values," pp. 50-52.

174. Wakefield, "Aristotle as Sociobiologist," pp. 25-28.

standing of what is meant by "reason" than Megone (or, arguably, Aristotle); so a good many of the functions that Wakefield thinks are *coordinated* by reason are understood by Megone as aspects *of* reason. With that caveat, Megone has no difficulty in agreeing on the connection between reason and human sociality. He denies, however, that this connection undermines his reading of Aristotle, namely, that reason is the natural function of the human kind.[175]

However, this dispute is related to another, for our purposes particularly significant, area of disagreement between Megone and Wakefield. This concerns the nature of teleological explanation, and includes a dispute about the relationship of "fact" and "value." Wakefield's critique here includes the following points:[176]

1. Megone's Aristotelian function argument is circular, because it uses function to determine the nature of the human good, and that cannot work if function is also defined *in terms* of the good.

2. Megone identifies functions as contributions to reproduction, but an additional term is needed to explain why these particular features of an organism (those that contribute to reproduction) are the ones identifiable as functions. In Megone's account, that additional term is that contributions to reproduction are "positively valued" (as Wakefield puts it) — they are identified as what a good member of the species does. However, it is not clear *why* contributions to reproduction are to be identified as good.

3. Wakefield offers various counterexamples and thought experiments (some more plausible than others) in support of the claim that "the 'ought' of evaluation cannot so easily be derived from the 'is' of reproductive contribution."[177] Wakefield's proposed alternative is his own evolutionary account of natural function, which (by contrast with his earlier papers, as we have seen) he now explicitly claims as teleological. He takes teleological explanation to be simply an unusual kind of efficient-causal explanation, in which a function is a distinctive kind of effect that can explain its own cause. The way this works (it will be recalled) is that an inherited biological trait has effects that enhanced the inclusive fitness of its

---

175. Megone, "Mental Illness, Human Function, and Values," p. 54.
176. Wakefield, "Aristotle as Sociobiologist," pp. 28-35.
177. One of the less plausible (p. 30) is the alien species in the film *Independence Day*, whose way of life depends on moving from planet to planet and destroying ecosystems. If this is to be understood as a natural function of that species (rather than a dysfunction or a distortion of a function), it is not clear that — beyond its superficial appeal as a plot device — it is a coherent scenario.

present bearers' ancestors, which resulted in its persistence and spread in the population.

4. Further examples are introduced of human artifacts (including the neutron bomb), projects (the Holocaust), and natural phenomena (including infectious disease) where recognizable functional or purposive design is not associated with positive evaluation.

5. On the strength of these objections, Wakefield argues that the Aristotelian claim that functions are characteristic of *good* members of a species can only be rescued by distinguishing between functional and evaluative senses of the word "good." So, referring to an earlier example of Megone's, he argues that a gardener who wishes to purchase a "good rose" is looking for one that is "good" in the functional sense, "that it is healthy and all of whose components will function as designed"; likewise, "[a] perfectly *good person,* in the sense of a person lacking in pathology, may not be a *good person* in the evaluative sense implying virtue or excellence."[178]

Megone's response to these points can be expressed (with some paraphrasing and reconstruction) as follows:[179]

1. Wakefield's first point misrepresents the Aristotelian argument. What Aristotle does is to locate the discussion of the human good in the broader context of a world that contains natural kinds whose functions are teleologically explicable, and this in turn informs a notion of how a good member of the *human* species can be expected to behave. As Megone persuasively remarks, "[t]he argument may suggest that certain concepts need to be understood together, but not in any manner that is viciously circular."[180]

2. In relation to Wakefield's second point, Megone acknowledges that it is not entirely clear from Aristotle's account *why* contributions to reproduction are to be identified as good.[181] Megone's proposal is that "the sense in which all this is good is given if we recognize that a stable, ordered, persisting ecosystem is better than a chaotic, degenerating ecological environment," because in any degenerating, chaotic environment *no* explanation in terms of natural function would be possible;[182] a stable persisting ecosystem is a *sine qua non* for any teleological explanation to be intelligible. But this remains a weak point in his account.

---

178. Wakefield, "Aristotle as Sociobiologist," p. 34, emphasis original.
179. Megone, "Mental Illness, Human Function, and Values," pp. 50-61.
180. Megone, "Mental Illness, Human Function, and Values," p. 56.
181. Megone, "Aristotle's Function Argument," p. 194.
182. Megone, "Mental Illness, Human Function, and Values," p. 57, referring back to "Aristotle's Function Argument," p. 194.

3. The third point goes to the heart of the difference between Megone and Wakefield. Megone draws attention to, and challenges, two key presuppositions that underpin this point; they also inform much of Wakefield's critique and positive account — and, Megone implies, some of the misconceptions identifiable in the latter.[183] The first is the fact-value separation; the second is that the only kind of causal explanation is efficient causation. Wakefield assumes both of these points without argument, but Megone argues that they cannot be taken for granted. He makes a case for a different view of the world informed by an Aristotelian fourfold scheme of causal explanation (material, formal, efficient, and final),[184] in which none of these modes of explanation is reducible to any of the others. Teleological explanation is not reducible to efficient cause, and is furthermore a mode of explanation that irreducibly involves the identification of some *good*.

Megone thus argues persuasively that an efficient-causal explanation which makes no reference to a goal identified as good is not a teleological explanation, even if it has the peculiar structure of an evolutionary explanation in which effects explain their causes. He also succeeds in showing (contra Wakefield) that the design of human *artifacts* cannot plausibly be explained without remainder in efficient-causal terms: the explanation of an artifact's design needs a teleological explanation that involves the identification of some good pursued by its designer.

In my judgment, however, he does not succeed in showing that a functional explanation of *biological structures and mechanisms* fails in the absence of a teleological account involving the identification of some good. He attempts to argue this by responding to Wakefield's use of the heart as an illustration of his evolutionary functional explanation: "the fact that past hearts had [the effect of pumping blood] causally explains how hearts came to exist and be maintained in the species and the genesis of the heart's detailed structure."[185] Megone responds:

> past hearts' pumping the blood figure in all sorts of causal stories, stories in which agents die young, or agents do not reproduce, or agents reproduce but defectively, and so on. It is not the case that hearts pumping the blood have simply caused hearts to exist . . .

---

183. Cf. "Mental Illness, Human Function, and Values," p. 59.

184. Cf. Aristotle, *Physics*, II.3, though, as Megone notes, the terminology of material, formal, efficient, and final causation is not Aristotle's; see further below, p. 145.

185. Wakefield, "Aristotle as Sociobiologist," p. 31.

> The best that might be said is that there is one causal chain ... in which, along with the operation of the rest of the body, does [sic] cause another heart to exist. But it is plausible to say that the only way to distinguish this causal chain from all the myriad others is to say that this is the chain in which the heart's pumping achieves its goal, survival of the species, which makes the pumping here thus derivatively good.[186]

This response misses an important point about evolutionary explanation: it is concerned with probabilistic effects operating across populations and over long periods of time. So a more careful way of formulating Wakefield's point is that hearts with a particular structure which enabled them to pump blood in a particular way would have the effect of making their bearers slightly more likely to survive and reproduce than the bearers of other kinds of heart, with the result that over many generations, this kind of heart became increasingly common in the population, until it was the only kind of heart found in members of this species. This is an efficient-causal explanation without the sort of gaps that require teleological filling. Because the cause and effect are averaged over a population, the explanation is not undermined if there are causal stories in which individuals died young or failed to reproduce; indeed, that is to be expected. Every individual who had one of those hearts rather than another kind thereby had an improved chance of surviving and reproducing; but a proportion of those individuals' lives could still be expected to be interrupted by some other (efficient) cause before they could reproduce. As long as this happened more rarely to bearers of these hearts than bearers of other kinds of heart, the causal explanation still works.

This is not to say that a teleological explanation cannot also be given; simply that an efficient-causal explanation of biological structures and their functions can be complete in its own terms. Modern biology does not need Aristotelian teleology to operate within its own frame of reference; it is simply that the story it tells about a biological organism such as a human being may not be the only story to be told.[187]

4. Wakefield's fourth point, it will be recalled, is that identifications of function or purpose can diverge from positive evaluations — as in the case of malign but effective political programs or weapons designed for wan-

---

186. Megone, "Mental Illness, Human Function, and Values," pp. 60-61.
187. See further, e.g., Messer, *Selfish Genes*, pp. 50-53.

tonly destructive purposes. Megone's response is quite simple: all that is required for a member of a natural kind, or its part, to be open to teleological explanation is that "there must be *some perspective* from which that activity is seen as good. Only from that perspective is the behavior functionally explicable."[188]

5. It will be clear from the responses to the preceding points that Megone does not consider Wakefield's proposed separation of functional from evaluative senses of "good" either warranted or necessary. Such a separation is made to seem both plausible and necessary by Wakefield's presuppositions that there is a fact-value distinction and that the only kind of causal explanation is efficient cause — both of which Megone rejects.

To sum up this debate: Megone's reading of Aristotle is more persuasive than Wakefield's. The latter's attempt to fit Aristotelian teleology onto the Procrustean bed of an evolutionary functional explanation is not really convincing; Megone makes a strong case for Aristotle's fourfold scheme of causal explanation and an understanding of teleological explanation which irreducibly involves the identification of some good. He is probably also correct to suggest that a properly teleological account, incorporating an understanding of the *good* of a member of the human species, is needed to support concepts of health and illness that can do the work required of them in relation to health care practice and bioethics. Bolting individually or socially determined evaluation onto a value-free scientific account of a human organism's structures and mechanisms, as Wakefield does in his "harmful dysfunction" model of disorder, is problematic for various reasons identified in earlier sections. If the evaluation of harm is determined by social conventions, then the understanding of health that follows seems vulnerable to serious distortion in toxic or unjust social environments. If the evaluation of harm is made by individuals in relation to whatever "vital goals" they happen to choose (as in Nordenfelt's holistic theory), the problem of individuals choosing unreal-

---

188. Megone, "Mental Illness, Human Function, and Values," p. 57. The theological resources to be introduced below in chapter 3 can go further than this in responding to Wakefield's counterexamples. In a Thomist understanding of teleology, any action of a rational agent — to be intelligible *as* an action — must be in pursuit of something that is identified as a good; but we are often mistaken in this identification, acting in pursuit of things that are not really goods, which is one way of understanding what the Christian tradition calls "sin." Furthermore, a theological account of "natural evil" — the insight that the world is God's good creation, but that not everything in the world as we experience it reflects God's good purposes — makes possible a response to Wakefield's "infectious disease" example.

istic, inappropriate or destructive goals persists, as critics of Nordenfelt have pointed out.[189]

However, Megone himself acknowledges the difficulty of explaining how the goods of natural kinds are recognizable as goods. Wakefield, as we saw, presses this point with various objections and counterexamples. Furthermore, Wakefield is quite correct to suggest that a modern biological account of an organism's function cannot supply an understanding of the organism's *good* in what he calls an evaluative sense, only (at most) in the non-evaluative "functional" sense in which cells being examined in a pathology laboratory might be described as "good" if they were free of lesions or damage and behaving as cells of that type were expected to behave. This is hardly surprising, because since early modernity the natural sciences have tended to proceed by systematically *excluding* final causes and goods from their explanatory schemata — a powerful and fruitful methodological step, provided it is not then taken to be a metaphysical claim.[190] To paraphrase a well-known remark of the British political spin-doctor Alistair Campbell, modern biology doesn't do value. If this reading is correct, then the impasse that we seem to have reached is that an understanding of the *good*, which must be an aspect of concepts of health, disease, and illness if they are to be capable of informing the practice of health care, cannot be supplied by the science that forms a large part of the theoretical base on which modern medicine is built.

An Aristotelian response to this problem could be to try and subject the same scientific data about the human being that are explicated in efficient-causal terms by a modern biological paradigm to a different kind of causal explanation, namely an explanation in terms of final causes. However, more work is needed to show whether and how the knowledge of human beings supplied by modern biological investigation is susceptible

189. See, e.g., Kovacs, "The Concept of Health and Disease."

190. The classic statement in favor of the methodological exclusion of teleology from the sciences is Francis Bacon, *The Advancement of Learning* 2, VII.2-7. Available in Francis Bacon, *The Advancement of Learning and The New Atlantis,* The World's Classics (London: Oxford University Press, 1913), pp. 99-107, and online at http://archive.org/details/advancementnewat00bacouoft. For a contrasting view of teleology in biology, see Michael Ruse, *Darwin and Design: Does Evolution Have a Purpose?* (Cambridge, MA: Harvard University Press, 2003), pp. 273-289. Ruse holds that neo-Darwinian evolutionary theory *is* teleological in something like an Aristotelian sense; as will be clear from the foregoing discussion, I agree with Megone against Ruse and Wakefield, taking the view that at most neo-Darwinism offers a *simulacrum* of genuinely teleological explanation.

to a teleological explanation that will yield a description of the human species as a natural kind (in an Aristotelian sense) with its own characteristic goods. This is, in effect, precisely the point at which Megone's presentation of Aristotle's function argument encountered some difficulty.[191] The danger is that unless some way can be found of showing that a description of this kind of being just *is* a description of a being whose good consists in certain characteristic things, Megone's identification of an organism's goals as good "from some perspective"[192] will collapse back into one of the forms of individually or socially constructed evaluation proposed in theories such as Wakefield's and Nordenfelt's.

There could be various ways out of this impasse, and it is quite possible that resources can be found within Megone's reading of Aristotle to resolve the problem. In any event, the theological account to be developed later in the book does offer a way around this impasse, by giving an account of human beings as *creatures* of a particular kind. It also offers a way of dealing with the problem cases adduced by Wakefield in an attempt to show that functional explanation does not necessarily go together with positive evaluation, such as infectious diseases, destructive artifacts and malign political programs. A theological account of sin and natural evil includes the resources to deal with these challenges in creative and fruitful ways.

Conclusion

To ask what is meant theologically by "health," "disease," and related concepts is to set up a highly interdisciplinary encounter with *(inter alia)* philosophical, scientific, and medical perspectives. If that encounter were structured so that its shape was determined by philosophical perspectives, which the distinctive sources and insights of the Christian tradition might then supplement or amend, then this would be the point at which to draw some detailed initial conclusions on the basis of the philosophical accounts and debates surveyed in this chapter. However, the approach of this book is in fact the reverse: to set up an engagement whose shape is determined by distinctively theological sources, methods, and content, into which insights and perspectives from other approaches and disciplines are

---

191. Megone, "Aristotle's Function Argument," p. 194.
192. E.g., Megone, "Mental Illness, Human Function, and Values," p. 57.

critically appropriated.[193] That being the case, it would be premature to draw detailed conclusions about the meaning of the health concepts at this stage of the account. What can be done at this stage, however, is to identify some of the insights and issues emerging from philosophical discussions of health with which it will be particularly important for the theological account developed in later chapters to engage.

1. The most obvious is teleology. A theological understanding of human beings as the creatures of a good and loving God is likely to prove receptive to a way of conceptualizing health that locates it within an understanding of the goods, goals, and ends of human creaturely life. A theological teleology will also have distinctive things to say about how the proper goals and ends of human life are to be identified and how we may recognize certain goals as *good*.

2. Since human creaturely life is *embodied* life, whose origins and structure are powerfully explained scientifically by evolutionary biology, the evolutionary account of biological functioning developed most fully by Jerome Wakefield will be an important area of understanding with which to engage. The uneasy relationship between an evolutionary understanding of this kind and the "value-laden" picture of human flourishing suggested by Aristotelian forms of teleology will be an important outstanding issue for a theological account to address — and, I have already suggested, one that it has the resources to deal with, creatively and fruitfully.

3. If human creaturely flourishing is understood in theological terms, then it seems promising to conceptualize health as the capacity to realize (some aspects of) that creaturely flourishing. For a theological account that understands health in something like this way — in Karl Barth's words, as "the strength for [human] life"[194] — Nordenfelt's account of abilities to realize vital goals and the kind of capability approach outlined by Law and Widdows might help to flesh out some of the language used by Barth to describe this "strength for life," such as "capability," "vigor," "freedom," and "integration." It must be emphasized again, though, that these insights should be *critically* appropriated, since there are aspects of them (such as Nordenfelt's want-satisfaction view of the happiness by which individuals' vital goals are set) that will be in tension with the theological account to be developed here.

4. I am skeptical that a clear-cut distinction can be drawn between "dis-

---

193. See below, pp. 103-107; and, further, Messer, *Selfish Genes*, pp. 49-61.
194. Barth, *CD* III.4, p. 356.

ease" (as a purely objective, scientifically describable pathological condition) and "illness" (as the subjective lived experience of loss of health) — essentially because of a theologically-motivated skepticism about the modern fact-value distinction that often underpins the disease-illness distinction. However, the difference — powerfully characterized by Toombs — between the perspectives of the clinician whose primary reality is the scientifically-understood "disease state" and the patient with his or her lived experience of illness, is one that cries out for a theological response in terms of both understanding and practice.

5. A sensitivity to the lived realities of human frailty, limitation, and suffering is oddly absent from some of the accounts surveyed, though impressively demonstrated in others, notably Toombs's phenomenological account. Broader questions about the meaning of suffering and evil are also present in the background of the debates surveyed in this chapter. For example, they are at least implicit in the exchanges between Wakefield and Megone about the divergence between the presence of functional design and its recognition as good both in human artifacts and in nature. A sensitivity to frailty, limitation, and suffering ought to lie at the very heart of Christian reflection and practice on health and disease, and — as suggested earlier — the central themes of the Christian theological tradition should have resources to offer to make sense of the puzzles concerning finitude, suffering, and evil that are encountered in the experience of illness and disease.

6. As has often been emphasized in the philosophical debates surveyed here (particularly, for example, by Fulford), an investigation of the meanings of health, disease, and illness has an essentially *practical* purpose. A theological account of health and disease will certainly emphasize this practical orientation, and aspects of Christian practice will be among the sources of the theological account to be developed in this book. What will be needed is a mutually critical encounter in which reflection on Christian practice informs the theological understanding of the questions and concepts being explored, but that practice is also challenged and informed by theological themes and insights, as well as by critical questions from other perspectives and disciplines.

In later chapters, I shall propose a theological approach that I believe will make it possible to appropriate these insights and address these issues. Before turning to that task, however, another area of critical discussion and questioning must be surveyed, namely the study of disability and the questions that it raises for theories of health, disease, and illness.

CHAPTER 2

Disability Perspectives: Critical Insights and Questions

Introduction

Since the 1960s, there have been major shifts in the understanding of disability, with important consequences for legislation, policy, and practice. Longstanding assumptions about disability have become contested; rival models and theoretical schemata have become the subject of vigorous (and sometimes bad-tempered) argument. Since there is clearly *some* kind of connection between what is said about disability and about health and disease — even if, in some forms of the social model, the connection only takes the form of a clear demarcation of boundaries — debates about disability inevitably raise critical questions for an understanding of health, disease, and illness. In this chapter I will survey selected debates about disability, and thereby identify the critical issues with which an account of health and disease needs to be properly engaged.

The chapter will begin with a fairly detailed comparison and contrast of two approaches to disability, the social model (drawing mainly, though not exclusively, on British disability studies literature) and what is sometimes called a socio-medical understanding (mainly represented here by the two successive World Health Organization classifications). The purpose of the comparison is to identify some of the most important disputed questions in recent debates about the conceptualization of disability, not to offer a comprehensive classification of disability models.[1] Next, a series

---

1. For a fuller and more systematic classification (though, as she acknowledges, still not complete), see Barbara M. Altman, "Disability Definitions, Models, Classification Schemes,

of connections will be explored between the approaches to disability surveyed in this chapter and the theories of health and disease discussed in the last. Finally, from the issues arising in these two sections, a set of critical questions will be identified to be taken forward into the development of a theological account of health.

A few comments on language, terminology, and perspective are in order. First, it should be noted that "disability" itself, as a unified category or object of study, is a relatively recent arrival on the scene. David Wasserman and his coauthors follow Ian Hacking in arguing that it only became possible to talk about "the disabled" as a unified group once categories of statistical normality and deviance had become established during the nineteenth century.[2] Furthermore, when social researchers began to study disability in the period following the Second World War, they quickly learned that it was a slippery and hard-to-define category — at least as much so as the categories of health, disease, and illness.[3] Secondly, there is an ongoing dispute about the appropriate language to use in discussing disability.[4] Some scholars and activists (more often in the United States than in Britain) favor the language of "people with disabilities," arguing that it puts people first and avoids defining them by their disabilities. Others (particularly in the British disability movement and disability studies community) favor the language of "disabled people" on the grounds that it signals that disability is a social disadvantage imposed on people through social, environmental, political, and economic factors. They argue that "people with disabilities" language im-

---

and Applications," in *Handbook of Disability Studies*, ed. Gary L. Albrecht, Katherine D. Seelman, and Michael Bury (Thousand Oaks, CA: Sage, 2001), pp. 97-122.

2. David Wasserman, Adrienne Asch, Jeffrey Blustein, and Daniel Putnam, "Disability: Definitions, Models, Experience," in *The Stanford Encyclopedia of Philosophy*, ed. Edward N. Zalta (Stanford: Stanford University Press, 2011), available online at http://plato.stanford.edu/archives/win2011/entries/disability/; Ian Hacking, *The Taming of Chance* (Cambridge: Cambridge University Press, 1990), pp. 160-169; see also Anita Silvers, "An Essay on Modeling: The Social Model of Disability," in *Philosophical Reflections on Disability*, ed. D. Christopher Ralston and Justin Ho (Dordrecht: Springer, 2010), pp. 19-36 (pp. 23-24).

3. Eda Topliss, *Provision for the Disabled*, 2nd ed. (Oxford: Blackwell, 1979), pp. 16-28; Altman, "Disability Definitions," p. 100.

4. See, e.g., Albrecht et al., *Handbook of Disability Studies*, p. 3; Colin Barnes, "Disability Studies: New or Not So New Directions?" *Disability and Society* 14.4 (1999): 577-580; Vic Finkelstein, "The Commonality of Disability," in *Disabling Barriers — Enabling Environments*, ed. John Swain, Vic Finkelstein, Sally French, and Mike Oliver (London: Sage, 1993), pp. 9-16; Tom Shakespeare, *Disability Rights and Wrongs* (London: Routledge, 2006), p. 24; Wasserman et al., "Disability," n. 1.

plies that disabilities are individual pathologies or conditions. In this book I follow the British convention, without wishing to imply thereby that disabled people are defined by disability. I also follow common British convention in using the term "impairments" to refer to individual differences or dysfunctions that may be associated with disability. A further terminological point concerns people with hearing impairments. Here I follow the widely observed convention of using the adjective "deaf" with a small "d" to refer in general to people with a hearing loss or impairment, and "Deaf" with a capital "D" to those deaf people who self-identify as members of a cultural and linguistic community defined in part by the use of a signed language.

In terms of perspective more generally, at the time of writing I would not identify myself as part of a "disability community," though of course the experience of disability has hardly been absent from my close family and immediate circle. I am acutely conscious of the disability movement's slogan, "Nothing about us without us."[5] While that slogan has sometimes been taken to mean that only disabled people are entitled to research and write about disability, it is increasingly argued in the disability literature that there are not two sharply differentiated categories of "the disabled" and "the non-disabled": any of us may find ourselves vulnerable, at some point in our lives, to impairments and the social harms that can be associated with them. I take the slogan to be primarily a protest against the labeling and stigmatization of disabled people, and against the description and theorizing of disability and the determination of policy and practice without proper attention to the experience and perspectives of disabled people themselves. Certainly it seems clear to me that one ought not to make theological claims about human flourishing, health, disease, and their implications for bioethics without attending carefully to those perspectives, and this chapter is an attempt to pay such attention.

Because the chapter has a specific purpose — to identify the critical insights and questions that a disability perspective can offer to the understanding of health and disease — it is not meant to be a comprehensive survey of the disability studies field. There are several aspects of the field — such as its implications for political theory and for detailed questions of social policy and practice — which, while not altogether absent from view, are not the primary focus and are not given the attention that a full treatment of the field would require.

---

5. James I. Charlton, *Nothing about Us without Us: Disability Oppression and Empowerment* (Berkeley: University of California Press, 2000).

Models and Theories of Disability

We turn first, then, to explore examples of two contrasting approaches to the understanding of disability: social and socio-medical models. Before beginning that exploration, it is worth noting a preliminary question about what is meant by calling these "models." Some scholars draw a sharp distinction between models and theories. For example, in response to criticisms of the alleged theoretical inadequacies of the social model, authors sometimes reply that it was never meant to be a theory.[6] But the distinction between models and theories in this area does not seem to be very clear-cut, particularly since some commentators deny that the so-called models are models at all. Anita Silvers, for example, understands a model to be "a standard, example, image, simplified representation, style, design, or pattern,"[7] and denies that the rival "models" of disability *could* be models in this sense. They are, however, used for the purposes to which models are often put, such as defining what disability is (and therefore also who counts as disabled), and explaining why it occurs. Barbara Altman notes other more specific purposes for which disability models have been created, such as information-gathering for social policy and epidemiology, understanding the problems associated with disability in order to identify solutions (in rehabilitation and other contexts), and drawing attention to neglected aspects of such environmental causes or influences on disability.[8]

It is clear that "models" of disability serve a range of purposes, and their design is partly determined by the work they are intended to do. This is not always innocent: as we shall see, social model authors complain that rival models serve ideological interests that are oppressive of disabled people, while critics of the social model retort that it too is ideologically-driven, in ways that can distort rather than aid understanding.[9] In any event, rival models do serve in this discussion as theoretical constructs that offer alternative ways of conceptualizing disability and human flourishing;

---

6. Michael Oliver, "The Social Model in Action: If I Had a Hammer," in *Implementing the Social Model of Disability: Theory and Research*, ed. Colin Barnes and Geof Mercer (Leeds: Disability Press, 2004), pp. 18-31 (pp. 23-24).

7. Silvers, "An Essay on Modeling," p. 22.

8. Altman, "Disability Definitions," pp. 111-112.

9. See, e.g., Mike Bury, "Defining and Researching Disability," in *Exploring the Divide: Illness and Disability*, ed. Colin Barnes and Geof Mercer (Leeds: Disability Press, 1996), pp. 17-38, and Mike Oliver, "Defining Impairment and Disability: Issues at Stake," in Barnes and Mercer, eds., *Exploring the Divide*, pp. 39-54.

this is not, however, an abstract theoretical exercise, because the understandings they offer have the most concrete of practical implications for health care and social and political life.

Social and Minority-Group Models

One of the most important recent shifts in the understanding of disability has been the development of "social" and "minority-group" models — the former more characteristic of British disability studies, the latter of the United States. While (as we shall see) there are significant differences between them, they also have important features in common — particularly their origins in disability activism and their emphasis on understanding disability in social rather than individual terms.

The origins of the British disability movement are usually traced to the 1960s and 1970s.[10] Campaigns on various disability issues led by the early 1970s to the formation of the Marxist-inspired Union of the Physically Impaired Against Segregation (UPIAS), as well as the more reformist Disability Alliance. UPIAS publications contain early expressions of what later came to be called the social model:

> In our view, it is society which disables physically impaired people. Disability is something imposed on top of our impairments by the way we are unnecessarily isolated and excluded from full participation in society. . . . To understand this it is necessary to grasp the distinction between the physical impairment and the social situation, called "disability," of people with such impairment. Thus we define impairment as lacking part of or all of a limb, or having a defective limb, organ, or mechanism of the body; and disability as the disadvantage or restriction of activity caused by a contemporary social organization which takes no or little account of people who have physical impairments and thus excludes them from participation in the mainstream of social activities. Physical disability is therefore a particular form of social oppression.[11]

---

10. See Vic Finkelstein, "A Personal Journey into Disability Politics," Centre for Disability Studies, University of Leeds, 2001. Available online at http://www.disability-archive.leeds.ac.uk/; and Shakespeare, *Disability Rights and Wrongs*, pp. 10-14.

11. Union of the Physically Impaired Against Segregation (hereafter UPIAS), *Fundamen-*

This statement expresses various key claims of the social model. It makes a sharp distinction between *impairments* (understood here as physical dysfunctions, later taken to include other forms such as sensory and cognitive impairments) and *disability* (a social situation imposed on people with impairments). It identifies disability as a form of oppression caused by the structure and organization of society, and implicitly denies a causal link between impairments and disabilities. The impairment/disability distinction has sometimes been even more sharply drawn, so that (for example) the Disabled People's International (DPI) could define disability without any reference to impairment, as "the loss or limitation of opportunities to take part in the normal life of the community due to physical or social barriers."[12]

Although these key ideas were gaining currency from the mid-1970s, the term "social model" itself did not emerge until slightly later, initially in the work of Michael Oliver.[13] The social model is often contrasted with a "medical model" of disability, and sometimes also with other models such as "welfare," "charity," and "administrative," but Oliver himself has stated a preference for a simpler contrast of *individual and social models,* in which "the idea underpinning the individual model [is] that of personal tragedy, while the idea behind the social model [is] that of externally imposed restriction."[14] The individual model will locate a disabled person's problem in their physical, cognitive, or sensory dysfunctions, which will be regarded as primarily medical problems calling for medical responses insofar as these are possible. David Wasserman and his co-authors observe that this model "is rarely defended but often adopted unreflectively by health care professionals, bioethicists, and philosophers who ignore or underestimate the contribution of social and other environmental factors to the limitations faced by people with disabilities."[15]

To illustrate the contrast between individual and social models, consider Barbara Altman's example of Jim.[16] Jim has cerebral palsy, which

---

tal *Principles of Disability* (London: UPIAS, 1976), p. 14. Available online at http://www.disability-archive.leeds.ac.uk/.

12. Disabled People's International, *Proceedings of the First World Congress, Singapore* (Ottawa: Disabled People's International, 1982), quoted in Shakespeare, *Disability Rights and Wrongs,* p. 14.
13. Mike Oliver, *Social Work with Disabled People* (Basingstoke: Macmillan, 1983).
14. Oliver, "The Social Model in Action," p. 19.
15. Wasserman et al., "Disability," section 2.
16. Altman, "Disability Definitions," p. 116.

does not affect his speech, but does limit many of his bodily functions. He uses an electric wheelchair, has a personal assistant who performs his personal care tasks, and a job coach who performs physical tasks that he cannot manage at his work in a hardware store (such as lifting and fetching items off shelves). In Oliver's schema, the individual model would locate Jim's problems and needs in his medical condition and the loss of bodily function that it has caused. The appropriate response would be clinical interventions to cure what dysfunctions can be cured and mitigate the effects of those that cannot. By contrast, the social model is "an attempt to switch the focus away from the functional limitations of individuals with an impairment on to the problems caused by disabling environments, barriers, and cultures."[17] In this model, the *disability* Jim experiences would be attributed not to the losses of function caused by his cerebral palsy, but to factors such as physical barriers to wheelchair access, the organization of a work environment that requires sales staff to lift heavy items and reach high shelves, and so on. The appropriate responses would be the forms of environmental, organizational, and political change needed to remove these disabling barriers.

Like the British social model, the American minority-group model of disability has its roots in campaigning and activism. David Braddock and Susan Parish identify four "watershed" developments in the civil rights of people with disabilities during the 1970s.[18] One was the passage of social security legislation which provided incentives to states to improve their service provision for people with cognitive impairments in line with federal standards, which had the effect of deinstitutionalizing care and service provision in this area. The second was a landmark legal judgment establishing a constitutional right to medical treatment for people with cognitive impairments in Alabama's institutions, which led to a large number of class actions relating to institutional conditions and access to education.[19] The third was legislation prohibiting discrimination against disabled people by groups or organizations receiving federal funding. Delays in the implementation of this legislation led to demonstrations and direct action, identified by Braddock and Parish as "one of the first times in American history that cross-disability advocacy groups had successfully worked to-

17. Oliver, "The Social Model in Action," p. 20.
18. David L. Braddock and Susan L. Parish, "An Institutional History of Disability," in Albrecht et al., *Handbook of Disability Studies*, pp. 11-68 (pp. 45-48).
19. Wyatt v. Stickney, 325 F. Supp. 781 (M. D. Ala. 1971), cited in Braddock and Parish, "An Institutional History of Disability," p. 47.

gether on a unified disability rights agenda."[20] This coalition-building had the effect of strengthening the disability rights movement and its campaigning for the wider-ranging changes later enshrined in the 1990 Americans with Disabilities Act (ADA). The fourth was the Education for All Handicapped Children Act (1975), which enshrined a right to publicly-funded education for disabled children and young people. According to Braddock and Parish, this had various effects, one being a shift from segregated towards integrated education for disabled children, another the creation of a generation of parents who became strongly involved in advocacy for the rights of disabled children. Other major foci of American disability campaigning and activism since the 1970s have been the independent living movement (which was an important influence on Paul Hunt, one of the founders of UPIAS in Britain) and the rise of self-advocacy groups for people with cognitive and sensory impairments and mental illnesses.[21]

Like its British counterpart, the American disability rights movement has emphasized the social causes of the problems faced by people with disabilities, and the need for social and political changes to address those problems.[22] However, as Tom Shakespeare notes, it is only relatively recently that the American literature has made much reference to the British social model.[23] While the British and American movements both emphasize the social factors in disability and the need for environmental, social, and political change to address those factors, there are important differences between the characteristic emphases of the two. The American movement has tended to understand people with disabilities as a minority group, and a major goal has accordingly been the passage of civil rights legislation to protect individuals in that group from unjust discrimination. Thus, the passage of the ADA in 1990 is seen as one of the movement's major achievements.[24] The British movement has been more influenced by the Marxist and materialist approach of UPIAS, and has understood itself as a grassroots movement working for wholesale political and social change. Prominent authors such as Paul Abberley and Michael Oliver have theorized disability in terms of the structure and operation of capitalist so-

---

20. Braddock and Parish, "An Institutional History of Disability," p. 47.
21. Braddock and Parish, "An Institutional History of Disability," pp. 48-50; for the influence of the independent living movement on Paul Hunt and UPIAS in Britain, see Finkelstein, "A Personal Journey," p. 3.
22. See Braddock and Parish, "An Institutional History of Disability," p. 44.
23. Shakespeare, *Disability Rights and Wrongs*, p. 24.
24. Braddock and Parish, "An Institutional History of Disability," p. 50.

ciety,[25] and have voiced skepticism about the benefits of British legislation such as the Disability Discrimination Act (1995).[26] Another point of contrast is that American literature and activism less commonly makes the sharp impairment/disability distinction characteristic of the British social model.[27] This contrast is related to the differences in the use of language already noted in the chapter introduction, particularly whether those who experience disability first-hand should be referred to as "disabled people" or "people with disabilities."

Critical discussion of the British social model in recent years has prompted various clarifications, modifications, and extensions, as well as proposals that it be abandoned and equally spirited defenses. This discussion and critique has raised various issues of importance for the present project.

First, there is a cluster of questions about *the connections between disease, impairment, and disability*. As we have seen, the British social model emphasizes the distinction between impairment and disability, contending that disability is caused by oppressive social structures and attitudes rather than impairment. One concern articulated in the social model is the inappropriate medicalization of disability, characterized by Oliver as an aspect of the ideology of individualism in capitalist societies which constructs the "disabled person" in a particular way:

> the disabled individual is an ideological construction related to the core ideology of individualism and the peripheral ideologies related to medicalization and normality. And the individual experience of disability is structured by the discursive practices which stem from these ideologies.[28]

---

25. Paul Abberley, "The Concept of Oppression and the Development of a Social Theory of Disability," *Disability, Handicap, and Society* 2.1 (1987): 5-19; Michael Oliver, *The Politics of Disablement* (London: Macmillan, 1990).

26. For British legislation, see *Disability Discrimination Act, 1995* c. 50 (hereafter DDA), available online at http://www.legislation.gov.uk/ukpga/1995/50/contents/enacted, and more recently *Equality Act, 2010* c. 15, available online at http://www.legislation.gov.uk/ukpga/2010/15/contents/enacted, which replaces most of the DDA. For skeptical comments on the DDA, see Mike Oliver and Colin Barnes, "Disability Politics and the Disability Movement in Britain: Where Did It All Go Wrong?" (June 2006), and Vic Finkelstein, "The 'Social Model of Disability' and the Disability Movement" (March 2007), both online at http://www.disability-archive.leeds.ac.uk/.

27. Shakespeare, *Disability Rights and Wrongs*, p. 24.

28. Oliver, *The Politics of Disablement*, p. 58.

However, the sharp separation between impairment and disability that the social model seems to promote has been criticized both by sociologists of health and by disability activists and academics. Among the former, Michael Bury writes in defense of the World Health Organization approach (to be discussed below) that "the idea that impairment and disability are closely related, but distinct, proves difficult to resist. Without some underlying initial problem, social responses would, so to speak, have nothing to respond to."[29] He argues that a social-model approach, defining disability simply in terms of social oppression, is reductionist and risks inhibiting research that could make a valuable contribution to understanding and practice.

Among activists, Liz Crow is one who has argued that more attention needs to be paid to impairment and its connection with disability. She is careful to emphasize the importance of the social model to her, as to the wider disability movement: it has been her "mainstay," enabling her to confront and overcome discrimination.[30] Yet she argues that the disabled people's movement, in its concern to resist individualist, personal-tragedy views of disability, has made the mistake of effectively denying the reality of impairment: it is safer not to mention that impairments can be bad things to have. This denial, according to Crow, has various unfortunate consequences. One is that it fails to acknowledge the lived experience of many disabled people:

> Impairment *is* problematic for many people who experience pain, illness, shortened lifespan, or other factors. . . . It is vital not to assume that they are experiencing a kind of false consciousness — that if all the external disabling barriers were removed they would no longer feel like this.[31]

Another danger is that denying the negative aspects of impairment — such as pain and fatigue — could lead the disability movement to structure its activities in ways that make it difficult for some disabled people to participate: "What we risk is a world which includes an 'elite' of people with impairments, but which for many more of us contains no real promise of

---

29. Bury, "Defining and Researching Disability," p. 30.
30. Liz Crow, "Including All of Our Lives: Renewing the Social Model of Disability," in Barnes and Mercer, eds., *Exploring the Divide*, pp. 55-73 (p. 56).
31. Crow, "Including All of Our Lives," p. 66, emphasis original.

civil rights, equality, or belonging."[32] A third is that the movement's resistance to the inappropriate medicalization of disability can be construed as a rejection of *all* forms of medical intervention, when in fact there is a need to explore what forms of medical intervention, such as pain relief, are helpful and needed.[33] Crow calls for a renewal of the social model to include a more nuanced appreciation of both objective and subjective aspects of impairment — a point taken up again below.

Crow is by no means alone in developing these criticisms: others who have made related points include Sally French and Jenny Morris.[34] Most authors who offer these critiques call for a more nuanced or developed social model that takes more account of the experience of impairment, though some — notably Tom Shakespeare — use them as part of an argument for the rejection of the social model and its replacement by a new and more adequate account.[35]

One response to these critiques is simply to resist the inclusion of the lived experience of impairment in the social model, on the grounds that "focusing on experiences rather than the causes of disability is the surest way to return to the confusion between impairment and disability that bedeviled the 'medical model of disability.'"[36] A difference of emphasis can be discerned in the response of Michael Oliver, who insists that the social model, properly understood, has never denied the negative aspects of impairment: its aim is simply to identify that component of disabled people's experience which can be changed by transforming unjust social structures and attitudes, and to work for positive change.[37] Nor does Oliver wish to deny the value of appropriate medical intervention. Indeed, in his 1990

---

32. Crow, "Including All of Our Lives," p. 60.

33. Crow, "Including All of Our Lives," pp. 65-66.

34. Sally French, "Disability, Impairment, or Something in Between?" in Swain et al., eds., *Disabling Barriers — Enabling Environments*, pp. 17-25; Jenny Morris, *Pride against Prejudice: Transforming Attitudes to Disability* (London: The Women's Press, 1991), p. 10.

35. Tom Shakespeare and Nicholas Watson, "The Social Model of Disability: An Outdated Ideology?" *Research in Social Science and Disability* 2 (2002): 9-28; Shakespeare, *Disability Rights and Wrongs*, pp. 38-43. See further below, p. 71.

36. Vic Finkelstein, "Outside, 'Inside Out,'" *Coalition* (April 1996): 30-36. Typescript available online at http://www.disability-archive.leeds.ac.uk/.

37. See, e.g., Oliver, "Defining Impairment and Disability," pp. 47-49; Oliver, "The Social Model in Action," pp. 21-23; Mike Oliver and Len Barton, "The Emerging Field of Disability Studies: A View from Britain," paper presented at Disability Studies: A Global Perspective (Washington, DC, October 2000), pp. 4-6, available online at http://www.disability-archive.leeds.ac.uk/.

analysis of the ideology of medicalization, he remarks that some medical interventions "are, of course, entirely appropriate, as in the diagnosis of impairment, the stabilization of medical condition after trauma, the treatment of illness occurring independent of disability and the provision of physical rehabilitation." His critique is directed at the extension of the medical profession's power into areas where he considers it less desirable, such as the assessment of a person's ability to drive or work, or involvement in the allocation of disability benefits.[38] Nonetheless, he is wary of calls for the social model to incorporate a fuller focus on the experience of impairment, for reasons similar to Finkelstein's: that to do so might encourage a lapse back into an individualist and biological understanding of disability.[39]

Another aspect of this discussion concerns the relationship between disease (chronic or otherwise) and impairment. Michael Bury, from a sociology of health perspective, makes a close connection between the two, citing with approval the understanding of impairment articulated in the first WHO classification: "abnormality in the structure [or] the functioning of the body, *whether through disease or trauma*."[40] By contrast, the social model expressly denies a causal relationship between disease and disability,[41] and some presentations even appear to deny or play down the connection between disease and *impairment*. For example, the DPI definition of impairment, "the functional limitation within the individual caused by physical, mental, or sensory impairment,"[42] is notoriously circular and makes no mention of disease or disorder. David Pfeiffer also questions the association between chronic illness and impairment, remarking that "many persons with disabilities have chronic *conditions* but are not ill."[43] Oliver, however, suggests that there may be more scope for rapprochement here than there appears:

38. Oliver, *The Politics of Disablement*, p. 48; see also David Pfeiffer, "The ICIDH and the Need for Its Revision," *Disability and Society* 13.4 (1998): 503-523 (pp. 509-510).

39. Oliver, "Defining Impairment and Disability," pp. 48-49.

40. Bury, "Defining and Researching Disability," p. 19, emphasis added, citing World Health Organization (WHO), *International Classification of Impairments, Disabilities and Handicaps* (Geneva: WHO, 1980; hereafter *ICIDH*). See also Ruth Pinder, "Sick-but-Fit or Fit-but-Sick? Ambiguity and Identity at the Workplace," in Barnes and Mercer, eds., *Exploring the Divide*, pp. 135-156.

41. See, e.g., Oliver, "Defining Impairment and Disability," pp. 41-42.

42. Disabled People's International, *Proceedings of the First World Congress*, quoted by Shakespeare, *Disability Rights and Wrongs*, p. 14.

43. Pfeiffer, "The ICIDH and the Need for Its Revision," p. 516.

It may well be that this debate is in reality, the result of terminological confusion; that real similarities exist between chronic illness and impairment and that there is much scope for collaboration between supporters of both schemas if this confusion can be sorted out.[44]

While, as we have seen, he is wary of attempts to develop a social model of impairment to complement the social model of disability, he allows that those disabled people wishing to develop such a model "may wish to develop a dialogue with medical sociologists working on the experience of chronic illness . . . our understandings of the experience of impairment may well be enhanced and the enterprise of medical sociology enriched."[45]

A second set of questions, roughly parallel to aspects of the discussion in the last chapter, is concerned with *what is objective and what subjective*, and *what is biological and what social*, in the concepts of impairment and disability. The impairment/disability distinction is sometimes compared to the sex/gender distinction in feminist theory: just as gender has often been understood as a social construction supervening on the objective reality of biological sex, so disability could be understood as a social construction supervening on the objective biological reality of impairment. Early disability definitions such as the UPIAS statement are read by critics such as Tom Shakespeare and Shelley Tremain as expressing this kind of binary opposition.[46]

Social-model authors, though, have not always understood impairment in such straightforwardly biological terms. As long ago as 1987, Paul Abberley argued for a "social theory of impairment" that would recognize social as well as biological causes and influences on impairment.[47] For example, in parts of the developing world, poverty and malnutrition are major causes of impairments such as blindness due to vitamin A deficiency. In industrialized countries, impairments due to injuries sustained at work or side-effects of prescribed drugs are among Abberley's examples of the influence of social context on patterns of impairment. Moreover, as Shakespeare points out, what counts as impairment is itself influenced by social context: dyslexia might not count as an impairment in a non-literate cul-

---

44. Oliver, "Defining Impairment and Disability," p. 42.
45. Oliver, "Defining Impairment and Disability," p. 52.
46. Shakespeare, *Disability Rights and Wrongs*, pp. 29-30, 34-38; Shelley Tremain, "On the Subject of Impairment," in *Disability/Postmodernity: Embodying Disability Theory*, ed. Mairian Corker and Tom Shakespeare (London: Continuum, 2002), pp. 32-47.
47. Abberley, "The Concept of Oppression," pp. 9-13.

ture.[48] For Abberley, this does not mean that impairment is socially constructed in the same sense as disability: a social view of impairment

> does not deny the significance of germs, genes, and trauma, but rather points out that their effects are only ever apparent in a real social and historical context, whose nature is determined by a complex interaction of material and nonmaterial factors.[49]

To use an example of his, there is a well-established material connection between smoking and diseases such as lung cancer; but social, cultural, and economic factors have a major influence both on patterns of tobacco consumption and on the medical services available to deal with its effects. For Oliver and others, such a social view of impairment is entirely consistent with the social model of disability.[50] Poststructuralist theorists, however, press the critique further. Shelley Tremain, for example, argues that like "sex," "impairment" should not be understood as a natural, pre-discursive phenomenon, but (in Michel Foucault's terms) as the outcome of a "disciplinary" exercise of "knowledge/power," which produces "subjects . . . who 'have' impairments because this identity meets certain requirements of contemporary political arrangements."[51] Like Shakespeare, Tremain holds that this analysis tends to dissolve the impairment/disability distinction: "In short, impairment has been disability all along."[52] This matters because the impairment/disability distinction was articulated in the social model to inform and motivate political action against the socially-constructed forms of oppression identified as disability. If impairment is indeed to be understood in Foucauldian terms as a regulatory category forming part of a disciplinary regime that renders its subjects docile,[53]

---

48. Shakespeare, *Disability Rights and Wrongs*, p. 35. Again, there are parallels with discussions of disease and disorder, such as the role ascribed to social context in Jerome Wakefield's "harmful dysfunction" model; see above, pp. 26-36.

49. Abberley, "The Concept of Oppression," p. 12.

50. E.g., Oliver and Barton, "The Emerging Field of Disability Studies," pp. 4-6.

51. Tremain, "On the Subject of Impairment," p. 42.

52. Tremain, "On the Subject of Impairment," p. 42; cf. Shakespeare, *Disability Rights and Wrongs*, p. 35.

53. Following Margrit Shildrick and Janet Price, Tremain gives the example of the questionnaire self-administered by applicants for the UK's Disability Living Allowance (DLA), which she interprets as an exercise in "self-surveillance," a "performance of textual confession [through which] the potential recipient is made a subject of impairment, in addition to being made a subject of the state, and is rendered docile"; "On the Subject of Impairment,"

then the potential of the social model to subvert oppressive power in relation to disability will be limited. "[I]f we continue to animate the regulatory fictions of 'impairment' and 'people with impairments' . . . the Disabled People's International (DPI) slogan, 'the right to live and be different,' will only translate as 'the right to live and be the same.'"[54]

In this connection, critics have also raised more general questions about the notions of biological normality that would underpin not only a biological understanding of impairment, but also many of the concepts of health and disease reviewed in the last chapter. This question will be taken up again later in the chapter.

A third area of critical discussion of the social model concerns its *coverage and representation:* whether it reflects the situation of all disabled people, or only a subset. One aspect of this critique is that women, ethnic minorities, and sexual minorities have been underrepresented in the British disability movement.[55] Social model authors such as Michael Oliver acknowledge some validity in this critique, but argue that there is nothing in the social model that makes this inevitable.[56] Disabled feminists such as Jenny Morris, however, hold that the underrepresentation of women's experience is not merely a contingent feature of the disability movement's history, but highlights a flaw in the social model itself: it has tended "to deny the experience of our own bodies, insisting that physical differences and restrictions are entirely socially created."[57] Feminism, she argues, can challenge social-model theorists to take fuller account of embodied experience:

> A feminist perspective on disability must focus, not just on the socioeconomic and ideological dimensions of our oppression, but also on what it feels like to be unable to walk, to be in pain, to be incontinent, to have fits, to be unable to converse, to be blind or deaf, to have an intellectual ability which is much below the average. There are posi-

---

p. 43, citing Margrit Shildrick and Janet Price, "Breaking the Boundaries of the Broken Body," *Body and Society* 2.4 (1996): 93-113.

54. Tremain, "On the Subject of Impairment," p. 45.

55. Shakespeare, *Disability Rights and Wrongs*, p. 13; Nasa Begum and Gerry Zarb, "Measuring Disabled People's Involvement in Local Planning," *Measuring Disablement in Society*, Working Paper 5 (1996), available online at http://www.disability-archive.leeds.ac.uk/; Morris, *Pride against Prejudice*, pp. 178-180.

56. Oliver, "Defining Impairment and Disability," p. 49; Oliver, "The Social Model in Action," p. 23.

57. Morris, *Pride against Prejudice*, p. 10.

tive and strong elements to these experiences but there are also negative and painful elements. The tendency of the disability movement to deny the difficult physical, emotional, and intellectual experiences that are sometimes part of the experience of disability is a denial of "weakness," of illness, of old age and death. . . . We should not be made to feel that we have to deny these negative things in order to assert that our lives have value.[58]

Other disabled feminists such as Carol Thomas and Mairian Corker have developed similar arguments.[59]

Another aspect of this critique is that the social model reflects the experience of people with some impairments better than others: it is said that in the early days of the British disability movement, UPIAS was composed predominantly of wheelchair users,[60] and it has been remarked that "[t]he social model of disability appears to have been constructed for healthy quadriplegics."[61] More specifically, various authors have reflected on the relative neglect of learning difficulties (or cognitive impairments) in the social model literature.[62] This seems to contrast with the American disability movement, in which people with cognitive impairments (and their families and caregivers) have had a higher profile from its early days.[63] However, it also points to a more general issue, raised by some of the theological authors on disability whose work will be reviewed in the next chapter, that the experience of cognitive impairments brings into focus the importance, but also the limits, of rights-based and politically activist responses to disability.[64]

---

58. Morris, *Pride against Prejudice*, pp. 70-71.

59. Carol Thomas and Mairian Corker, "A Journey around the Social Model," in Corker and Shakespeare, eds., *Disability/Postmodernity*, pp. 18-31; Carol Thomas, "Developing the Social Relational in the Social Model of Disability: a Theoretical Agenda," in Barnes and Mercer, eds., *Implementing the Social Model of Disability*, pp. 32-47.

60. Finkelstein, "A Personal Journey," p. 4.

61. R. Humphrey, "Thoughts on Disability Arts," *Disability Arts Magazine* 4.1 (1994): 66-67 (p. 66), quoted by Oliver, "Defining Impairment and Disability," p. 49.

62. See, e.g., Anne Louise Chappell, "From Normalization to Where?" in *Disability Studies: Past, Present, and Future*, ed. Len Barton and Mike Oliver (Leeds: The Disability Press, 1997), pp. 45-61; Jan Walmsley, "Including People with Learning Difficulties: Theory and Practice," in Barton and Oliver, eds., *Disability Studies*, pp. 62-77; Shakespeare, *Disability Rights and Wrongs*, pp. 74-76.

63. See Braddock and Parish, "An Institutional History of Disability," pp. 44-50.

64. See, e.g., Medi Ann Volpe, "Irresponsible Love: Rethinking Intellectual Disability, Humanity, and the Church," *Modern Theology* 25.3 (2009): 491-501.

Socio-Medical Models

Since hardly anyone actually argues for an extreme "medical model" of the sort that social-model theorists characterize — however much it may be practiced *unreflectively*, as David Wasserman and his coauthors suggest[65] — the chief alternatives to the social model are what Michael Bury calls "socio-medical models": those that combine medical, social, and other factors in an account of disablement.[66]

Among the most influential models in this category — and those that have provoked the most argument and criticism — are the two successive versions published by the World Health Organization.[67] The first of these, *ICIDH*, was developed in order to meet various perceived needs, including WHO's requirement to gather information about nonfatal health outcomes in order to gain a fuller international picture of the health status of populations, and the need for a clear conceptual framework to inform social research and policymaking concerning disability.[68] The resulting conceptual scheme comprises four elements: disease, impairment, disability, and handicap. Their relationships are conceptualized as follows:

> (i) *Something abnormal occurs within the individual:* this may be present at birth or acquired later. A chain of causal circumstances, the "etiology," gives rise to changes in the structure or functioning of the body, the "pathology" . . .
>
> (ii) *Someone becomes aware of such an occurrence:* in other words, the pathological state is exteriorized. . . . In behavioral terms, the individual has become or been made aware that he is unhealthy. . . . *In the*

---

65. Wasserman et al., "Disability."

66. Bury, "Defining and Researching Disability"; Michael Bury, *Health and Illness in a Changing Society* (London: Routledge, 1997), ch. 4.

67. WHO, *ICIDH*; WHO, *International Classification of Functioning, Disability, and Health* (Geneva: WHO, 2000; hereafter *ICF*).

68. Bury, "Defining and Researching Disability," citing M. Jefferys, J. B. Nullard, M. Hyman, and M. D. Warren, "A Set of Tests for Measuring Motor Impairment in Prevalence Studies," *Journal of Chronic Diseases* 28 (1969): 303-309; A. Harris, E. Cox, and C. Smith, *Handicapped and Impaired in Great Britain*, vol. 1 (London: HMSO, 1971), and D. Patrick and H. Peach, eds., *Disablement in the Community* (Oxford: Oxford Medical Publications, 1989). See also T. B. Üstün, S. Chatterji, J. Bickenbach, N. Kostanjsek, and M. Schneider, "The International Classification of Functioning, Disability, and Health: A New Tool for Understanding Disability and Health," *Disability and Rehabilitation* 25.11-12 (2003): 565-571 (esp. pp. 565-566).

context of health experience, an impairment is any loss or abnormality of psychological, physiological, or anatomical structure or function.

(iii) *The performance or behavior of the individual may be altered as a result of this awareness, either consequentially or cognitively. Common activities may become restricted, and in this way the experience is objectified. . . . In the context of health experience, a disability is any restriction or lack (resulting from an impairment) of ability to perform an activity in the manner or within the range considered normal for a human being.*

(iv) *Either the awareness itself, or the altered behavior or performance to which this gives rise,* may place the individual at a disadvantage relative to others, thus socializing the experience. . . . *In the context of health experience, a handicap is a disadvantage for a given individual, resulting from an impairment or a disability, that limits or prevents the fulfillment of a role that is normal (depending on age, sex, and social and cultural factors) for that individual.*[69]

The contrasts between these definitions and those of the social model are obvious. In *ICIDH,* impairment is explicitly linked to disease or pathology and identified as the cause of disability. By contrast to the social model, disability is here defined as the individual's lack of functional capacity. "Handicap," not "disability," is the term used to conceptualize the social aspect of disablement.

ICIDH was widely used and influential, being adopted by a number of European countries, U.S. states, academic institutions, and international organizations for purposes that included data collection, epidemiological research, social policy, planning of service delivery, and assessment of disabled people for employment and welfare purposes.[70] However, although the *ICIDH* schema was explicitly intended to articulate an understanding of disablement as a social phenomenon and to move away from an exclusively medical and individual understanding, it was severely criticized by disability activists and social-model authors, who interpreted it as essentially a version of the individual or medical model of disability. Moreover, social-model advocates were not the only critics of *ICIDH:* it was also

69. WHO, *ICIDH,* pp. 25-29 (emphases original), quoted by Saad Z. Nagi, "Disability Concepts Revisited: Implications for Prevention," in *Disability in America: Toward a National Agenda for Prevention,* ed. Andrew M. Pope and Alvin R. Tarlov (Washington, DC: National Academies Press, 1991), pp. 309-327 (pp. 319-320).

70. Pfeiffer, "The ICIDH and the Need for Its Revision," p. 505.

found wanting in some respects by some social researchers, medical organizations, and authors of other socio-medical models. Among the main criticisms leveled at it were the following:

1. Disabled people and disabled people's organizations themselves were not involved in drawing up the definitions in *ICIDH*, with the result that the perspectives of the health care professions rather than disabled people themselves were privileged in defining disability and its effects.[71]

2. The *ICIDH* schema locates the cause of disablement in diseases, disorders, or pathologies suffered by individuals, precisely the causal connection that the social model denies.[72] As such it has been read as medicalizing disability, attributing the social discrimination suffered by disabled people to their individual impairments — a move that Pfeiffer understands as stigmatizing and blaming the victim in ways analogous to those that have operated in relation to other forms of social oppression such as racism and sexism.[73]

3. A related criticism concerns the emphasis placed on *normality* in the *ICIDH* definitions of both disability and handicap. Various critics have argued that the concept of the "normal" is distinctly slippery and its uses in health care involve a good deal of equivocation: it can refer to a statistically typical level or range of function, the absence of a disease state, an acceptable state or behavior, and sometimes even an ideal state.[74] This slippery concept, it is argued, is deployed in the stigmatizing and victim-blaming ways already noted so that the disadvantages suffered by disabled people are laid at the door of their "abnormalities" rather than the hostile social environments in which they find themselves. In Ron Amundson's words, "[p]hilosophers and medical practitioners alike have used the category [of normality] to conclude that the disadvantages of disabled people result

---

71. Oliver, *Politics of Disablement*, pp. 4-6; Rachel Hurst, "To Revise or Not to Revise?" *Disability and Society* 15.7 (2000): 1083-1087 (p. 1083). For a riposte to a version of this critique leveled at disability research in general, see Bury, "Defining and Researching Disability," p. 28.

72. Oliver, "Defining Impairment and Disability," pp. 41-42; see further Rachel Hurst, "The International Disability Rights Movement and the ICF," *Disability and Rehabilitation* 25.11-12 (2003): 572-576 (pp. 573-574).

73. Pfeiffer, "The ICIDH and the Need for Its Revision," pp. 513-515.

74. Anita Silvers, "A Fatal Attraction to Normalizing: Treating Disabilities as Deviations from 'Species-Typical' Functioning," in *Enhancing Human Traits: Ethical and Social Implications*, ed. Erik Parens (Washington, DC: Georgetown University Press, 1998), pp. 95-123 (pp. 104-105), quoting Philip Davis and John Bradley, "The Meaning of Normal," *Perspectives in Biology and Medicine* 40.1 (1996): 68-77 (pp. 69-70).

from their own abnormality; they have only themselves (and nature) to blame."[75] Furthermore, the stipulation that disability is lack of ability "to perform an activity *in the manner* or within the range considered normal . . ." is held to reflect a set of assumptions about the desirability of "normalization" that often work to the disadvantage of disabled people. For example, in some places and times, deaf children have been encouraged or coerced into lip-reading and speaking rather than using sign-language, because the former appears a more normal mode of communication, even when sign language could have enabled them to function at a higher level, and children with missing or shortened lower limbs have been coerced into using uncomfortable and only partially effective prostheses when wheelchairs would have given them greater mobility and comfort.[76]

4. A further critique is that the *ICIDH* model reflects an assumption that disabled people have a low quality of life (QoL) by virtue of their disabilities. Pfeiffer presses this critique to the point of arguing that *ICIDH* "facilitates the type of thinking which leads us quickly to Eugenics and the development of the master race,"[77] a charge that Michael Bury dismisses as "spurious."[78] While it is true that the passages quoted by Pfeiffer from *ICIDH* do not directly make the connection with judgments and assumptions about low QoL, it is also the case that this connection is frequently made. For example, as Pfeiffer notes, a judgment that wheelchair use — however adept — represents a lower quality of life than even restricted walking is built into standard QoL assessment instruments.[79] Furthermore, a number of studies have shown that non-disabled members of the public and health professionals tend to estimate the QoL of disabled people as very low, whereas disabled people tend to rate their own QoL far more highly. This discrepancy is sometimes rationalized as a form of self-

---

75. Ron Amundson, "Against Normal Function," *Studies in History and Philosophy of Biological and Biomedical Sciences* 31.1 (2000): 33-53 (p. 51).

76. Silvers, "A Fatal Attraction to Normalizing," pp. 110-115. See also Janet Price and Margrit Shildrick, "Bodies Together: Touch, Ethics, and Disability," in Corker and Shakespeare, eds., *Disability/Postmodernity*, pp. 63-75 (pp. 67-68), quoting Mary Duffy, "Making Choices," in *Mustn't Grumble: Writing by Disabled Women*, ed. L. Keith (London: Women's Press, 1994).

77. Pfeiffer, "The ICIDH and the Need for Its Revision," p. 510.

78. Mike Bury, "A Comment on the ICIDH2," *Disability and Society* 15.7 (2000): 1073-1077 (p. 1075).

79. Pfeiffer, "The ICIDH and the Need for Its Revision," p. 512; see, e.g., Health Utilities Index Mark 3 (HUI3), available online at http://www.healthutilities.com/.

deception on the part of disabled people, an interpretation robustly challenged by authors such as Amundson.[80]

5. Social researchers and others attempting to use and operationalize the *ICIDH* schema have identified various problems. To give a few examples: (i) the categories of impairment, disability, and handicap are not clearly differentiated from one another — for instance, both "disability" and "handicap," as they are operationalized in the classification, turn out to include both individual and social components; (ii) some of the categories appear to mix up descriptions of conditions or states with possible explanations of those states; (iii) the conceptual scheme is not a good fit to the real world — it fails to "[carve] at the joints."[81]

6. The use of the term "handicap" has been criticized as stigmatizing and offensive to disabled people.[82]

These were some of the factors that motivated a radical revision of *ICIDH*, culminating in the publication in 2001 of the *International Classification of Functioning, Disability, and Health (ICF)*.[83] *ICF* offers a radically different model of disability from that of *ICIDH*: disability is now understood as arising from a dynamic interaction between *health conditions* (diseases, disorders, injuries, etc.) and *contextual* (environmental and personal) *factors*. This interaction has two components: first, body functions and structures, and secondly, activities and participation. Impairments are understood as losses or abnormalities of body functions and structures, "activity limitation" replaces the category of "disability" in *ICIDH*, and "participation restriction" replaces "handicap." "Functioning" and "disability" are now umbrella terms for the positive and negative aspects respectively of the whole interaction between the individual and their context, which may involve either or both components.[84]

ICF can be said to address many of the criticisms leveled at *ICIDH*, at

---

80. Amundson, "Against Normal Function," p. 46; Ron Amundson, "Quality of Life, Disability, and Hedonic Psychology," *Journal for the Theory of Social Behavior* 40.4 (2010): 374-392; Ian Basnett, "Health Care Professionals and Their Attitudes toward and Decisions Affecting Disabled People," in Albrecht et al., *Handbook of Disability Studies*, pp. 450-467.

81. Bury, "Defining and Researching Disability," p. 20; Nagi, "Disability Concepts Revisited," pp. 320-327, quoting A. Kaplan, *The Conduct of Inquiry* (San Francisco: Chandler, 1964).

82. Pope and Tarlov, *Disability in America*, p. 6.

83. World Health Organization, *International Classification of Functioning, Disability, and Health*, short version (Geneva: World Health Organization, 2001; hereafter *ICF*).

84. WHO, *ICF*, pp. 190-191.

least up to a point. In relation to the numbered points listed earlier, the following comments may be made.

1. The revision process did involve extensive consultation, including disabled people and disability organizations, with the result that there is less dominance of health care professionals' perspectives in defining and conceptualizing disability in *ICF*.[85]

2. The model of disability that informs *ICF* is described as a "biopsychosocial model"; its authors claim that this model captures the important insights of both medical and social models while avoiding the reductionism of either.[86] Unlike *ICIDH*, *ICF* explicitly refrains from making any general claims about causality. The model suggests that the different factors and components can interact in a wide variety of ways, and it is left to users of the model to determine the particular causal relationships operating in individual cases.[87] Whereas the understanding of *ICIDH* could be schematized as follows:

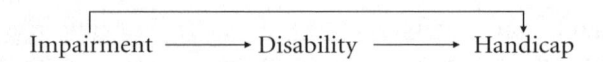

Impairment ⟶ Disability ⟶ Handicap

the *ICF* schema is more complex and open:

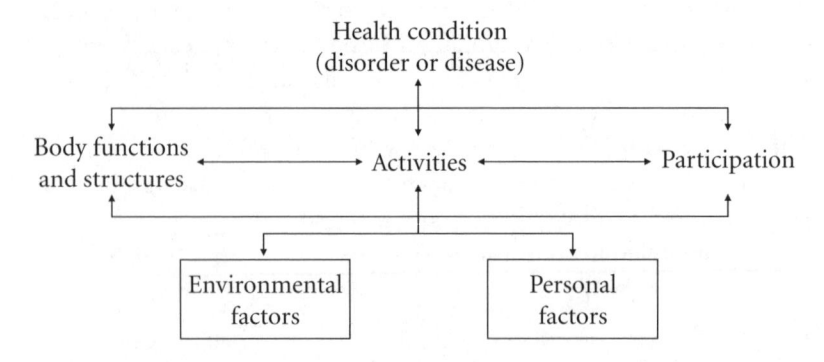

Health condition
(disorder or disease)

Body functions and structures ⟷ Activities ⟷ Participation

Environmental factors   Personal factors

(Redrawn from WHO, *ICF*, p. 26)

---

85. Hurst, "To Revise or Not to Revise?" pp. 1084-1085; Hurst, "The International Disability Rights Movement and the ICF."

86. WHO, *ICF*, p. 28; Üstün et al., "ICF: A New Tool," pp. 567-68. Rachel Hurst, however, argues that this misrepresents the social model, "which is, in fact, exactly the same as the ICF's 'interactive' model" ("The International Disability Rights Movement and the ICF," p. 575)!

87. WHO, *ICF*, p. 5.

Marguerite Schneidert and her colleagues, for example, give a range of examples to show how *ICF* can shift the focus away from individuals' impairments and identify the social and environmental changes needed to address various kinds of disablement.[88] Others, however, have questioned the completeness of the *ICF* schema. For example, Ueda and Okawa argue that it fails to represent the subjective experience of disability adequately, and they propose an extension of the model to incorporate subjective aspects — though arguably a good deal of what they have in mind is in principle included in the "personal factors" component of *ICF*.[89] The importance of these factors in disability is explicitly acknowledged, although they are not coded in the classification — presumably because they were considered too diverse, complex, and context-specific to be represented adequately in a usable standardized classification.

3. *ICF* has moved away somewhat from the concept of normality deployed in *ICIDH*, and in particular has abandoned the stipulation about the normal manner as well as range of functional performance in the latter's definition of disability. However, notions of normality — albeit "softer" than in *ICIDH* — are still in evidence in the *ICF* definitions of impairment, activity limitation, and participation restriction. The definition of impairment includes both a criterion of statistical normality and an assertion of professional authority in its assessment:

> Impairments represent a deviation from certain generally accepted population standards in the biomedical status of the body and its functions, and definition of their constituents is undertaken primarily by those qualified to judge physical and mental functioning according to these standards.[90]

Again, in the activity and participation components, two kinds of comparison are envisaged: first, a person's *performance* ("what individuals do in their current environment") is compared to their *capacity* ("the highest

---

88. Marguerite Schneidert, Rachel Hurst, Janice Miller, and Bedirhan Üstün, "The Role of Environment in the International Classification of Functioning, Disability, and Health (ICF)," *Disability and Rehabilitation* 25.11-12 (2003): 588-595.

89. S. Ueda and Y. Okawa, "The Subjective Dimension of Functioning and Disability: What Is It and What Is It For?" *Disability and Rehabilitation* 25.11-12 (2003): 596-601; cf. WHO, *ICF*, pp. 23-24.

90. WHO, *ICF*, p. 16. This definition is glossed on p. 190 as "a deviation from a population mean within measured standard norms."

probable level of functioning that a person may reach in a domain in the Activity and Participation list at a given moment . . . measured in a uniform or standard environment").[91] This gives a measure of the extent to which their present environment enables or limits their activity and participation, reflecting one of the key insights of the social model. Secondly, however, activity limitations and participation restrictions "are assessed against a generally accepted population standard," namely "that of an individual without a similar health condition."[92]

4. There is little evidence in *ICF* itself of problematic assumptions about the quality of life of disabled people, and also a clear concern to avoid segregating a subset of the population labeled "disabled people":

> There is a widely held misunderstanding that ICF is only about people with disabilities; in fact, it is about *all people.* The health and health-related states associated with all health conditions can be described using ICF. In other words, ICF has universal application.[93]

However, critics argue that questionable assumptions about QoL are still in evidence more broadly within WHO, and influence the ways in which *ICF* is used. In particular, Rachel Hurst notes the increasing use of disability-adjusted life years (DALYs) as a measure of nations' health status, interpreting this to mean that "we still have within WHO a system operating on the medical and individual model of disability alongside the interactive and rights based definition of the ICF and its environmental factors."[94]

5. Some of the conceptual anomalies and ambiguities discerned in *ICIDH* by its critics have disappeared with the introduction of the new categories in *ICF,* though some commentators have complained that new conceptual problems arise in the latter.[95] For example (as already noted), the "Activity" and "Participation" components have two qualifiers: *performance,* which assesses how an individual actually performs in their current environment, and *capacity,* which is related to a standardized environment.[96] Lennart Nordenfelt has questioned the use of a performance quali-

91. WHO, *ICF,* p. 192.
92. WHO, *ICF,* p. 21.
93. WHO, *ICF,* p. 8.
94. Hurst, "The International Disability Rights Movement and the ICF," p. 575.
95. Hurst, "To Revise or Not to Revise?" p. 1085.
96. WHO, *ICF,* pp. 19-20.

fier, which assesses what a person actually does rather than what they are capable of doing: put in the terms of his version of action theory, if a person fails to do something, this could be a reflection of their *will* rather than their ability or opportunity.[97] In other words, their lack of performance might be simply because they choose not to do something, not because they are prevented by internal limitations or external barriers, and if so, their lack of performance will not be informative about their functioning or disability. In place of the performance qualifier, Nordenfelt proposes an "opportunity qualifier" to assess what he calls "nonstandard situations" — for example, when "a person who is basically healthy and wants to do something is prevented from doing this by some salient obstacle," or in those instances when "[t]he provision of opportunities not only for people with normal capabilities, but also and indeed in particular, for people with subnormal capabilities" is the proper business of health or rehabilitation services.[98]

Nordenfelt is no doubt theoretically correct: lack of performance could in some cases indicate a lack of will that has no significance for the assessment of those individuals' disablement. However, his objection appears to miss one of the points of the performance and capacity qualifiers, namely that the comparison of a person's actual performance in their present environment with their capacity in a standardized environment offers a way of identifying barriers in their actual environment that might need to be dealt with. It is not clear that an "opportunity qualifier" would enable this identification to be made as clearly or directly in practical situations, even if it is conceptually tidier.[99]

Another of Nordenfelt's objections to the "present environment" stipulation in the *ICF* performance qualifier is instructive in this connection. He remarks that his own theory of health relates individuals' abilities to standard circumstances, because

> [t]he ability entailed in health cannot in general be looked upon from the point of view of the present environment. The reason is that the

---

97. Lennart Nordenfelt, "On Health, Ability, and Activity: Comments on Some Basic Notions in the ICF," *Disability and Rehabilitation* 28.23 (2006): 1461-1465. For a systematic presentation of disability in terms of Nordenfelt's action theory, see Lennart Nordenfelt, "Ability, Competence, and Qualification: Fundamental Concepts in the Philosophy of Disability," in Ralston and Ho, eds., *Philosophical Reflections on Disability*, pp. 37-54.

98. Nordenfelt, "On Health, Ability, and Activity," p. 1464.

99. Cf. Kath McPherson, "What Are the Boundaries of Health and Functioning — and Who Should Say What They Are?" *Disability and Rehabilitation* 28.23 (2006): 1473-1474.

> present environment can be extremely harsh or extremely demanding. In such extreme situations also healthy people may fail to act. However, most "present" environments actually fall within the standard, as I see it. The healthy person should be able to handle situations within a wide range of circumstances. . . . Therefore, if a person is unable to perform a particular action in the present situation he or she is for the most part . . . unhealthy, according to my analysis.[100]

Surely, however, one point of the "present environment" stipulation is that there are many ways in which quite unexceptional "present" environments prove disabling to some individuals. In Nordenfelt's account, this would reflect a lack of health in the individuals concerned, since health is defined in terms of ability to function in standard environments. Yet this conceptual connection between (lack of) health and disability would be strongly contested by the social model, and is one on which the authors of *ICF*, informed by that model, were careful to remain neutral. There is, in other words, an element of circularity in Nordenfelt's critique of *ICF*. In effect, he stipulates a basically individual definition of disability, excluding the central claim of the social model from the outset. He seems to take this understanding of disability, and the account of the relationship between health, disease, and disability that it implies, as axiomatic and to criticize *ICF* for not giving a coherent version of that account[101] — whereas the authors of *ICF*, informed in part by the social model, deliberately avoided stipulating any particular account of that relationship.

All of this raises a more general question about the relationship between models of disability and theories of health, which will be taken up again in the next section.

6. The term "handicap" has been abandoned, and in general the terminology is intended to be more positive and less focused on deficit or loss: the change is described as a move "away from being a 'consequence of disease' classification (1980 version) to become a 'components of health' classification."[102]

Some disability activists and social-model advocates have remained skeptical of the *ICIDH* revision process and hostile to *ICF*.[103] Others, how-

---

100. Nordenfelt, "On Health, Ability, and Activity," p. 1464.
101. Nordenfelt, "On Health, Ability, and Activity."
102. WHO, *ICF*, p. 5.
103. E.g., David Pfeiffer, "The Devils Are in the Details: The ICIDH2 and the Disability Movement," *Disability and Society* 15.7 (2000): 1079-1082.

ever, have cautiously welcomed it, while still maintaining that it has significant flaws.[104] This perhaps reflects a wider, if partial, *rapprochement* between the two broad approaches to disability surveyed in this section. The influence of feminist, poststructuralist, and phenomenological perspectives has led some authors and activists working within the social-model paradigm to place greater emphasis on the embodied experience of chronic illness and impairment, while professional communities in health care have become more ready to recognize the social dimensions of disability. As Gareth Williams puts it,

> [w]hile the medical model has become more social in response to social criticism, political protest, and the limits of medical interventions, the social model has become less unitary as it has been exposed to debate inside and outside disability studies.[105]

Williams himself endorses these attempts "to develop more systematically pluralistic approaches to disability and break away from . . . a theoretically sterile and rather contrived distinction between a social model and a medical model."[106] He is not alone in this. Tom Shakespeare, whose earlier work in disability studies was clearly located within the British social-model paradigm, has become increasingly critical of it, to the point of arguing in his more recent publications that it is best abandoned. He argues instead for an understanding of disability as a complex interaction "between factors intrinsic to the individual, and extrinsic factors arising from the wider context in which she finds herself"[107] — which, as he recognizes, is exactly the kind of understanding expressed in *ICF*.[108] Furthermore, there are others, such as Anita Silvers, who remain more positive about the social model and wish to defend it, but nonetheless hold that

---

104. E.g., Hurst, "To Revise or Not to Revise?" and "The International Disability Rights Movement and the ICF."

105. Gareth Williams, "Theorizing Disability," in Albrecht et al., eds., *Handbook of Disability Studies*, pp. 123-144 (p. 136).

106. Williams, "Theorizing Disability," p. 125.

107. Tom Shakespeare, "Disability: Suffering, Social Oppression, or Complex Predicament?" in *The Contingent Nature of Life: Bioethics and the Limits of Human Existence*, ed. Marcus Düwell, Christoph Rehmann-Sutter, and Dietmar Mieth (Dordrecht: Springer, 2008), pp. 235-246 (p. 241).

108. Shakespeare, *Disability Rights and Wrongs*, pp. 59-60.

In a pluralistic society, we should expect that different models of disability will be appropriate to realize different values, and that these will be as compatible, or as antithetical, as the values they serve. . . . For a pluralistic society, many models of disability are better than one.[109]

It is not the main purpose of this chapter to propose or advocate one approach to disability over others, though I do find proposals such as Shakespeare's "interactional" model persuasive. The aim of this survey is rather to identify, from the various approaches to disability and the arguments between their proponents, critical perspectives and questions that can be brought to the project of this book, which is to develop a theological account of health and disease. Accordingly, in the next section, I shall explore some connections between the accounts of disability surveyed in this chapter and the theories of health discussed in the last. The final section will then identify some critical questions to be taken forward into the later chapters where a constructive theological account will be developed.

Disability and Health

This section is an attempt to clarify the boundaries and connections between the accounts of health, disease, and illness discussed in the previous chapter and the accounts of disability surveyed in the present one. On one reading of the social model of disability, it might appear as though all that need be done on that score is to define the border between health concepts and disability clearly and police it carefully. Some accounts could reinforce that impression: for example, in his influential text *Understanding Disability*, Michael Oliver draws a sharp contrast by remarking that "[m]ost illnesses are treatable and even curable by medical interventions; most impairments are not curable; and all disability can be eradicated by changes to the way we organize society."[110] This is echoed by the sharp distinction sometimes made by social-model authors between chronic illness and impairment, as noted earlier.[111] However, Oliver himself suggests that these

---

109. Silvers, "An Essay on Modeling," p. 34.
110. Michael Oliver, *Understanding Disability: From Theory to Practice*, 2nd ed. (Basingstoke: Palgrave Macmillan, 2009), p. 44.
111. For a brief discussion, see Shakespeare, *Disability Rights and Wrongs*, p. 60.

distinctions are not a denial of all relationship between the two areas,[112] and as we saw in the last section, those working within a social-model paradigm have become more diverse in recent years in their views of the relationships between illness, impairment, and disability. In what follows, it will be assumed that there *is* a relationship to be explored, which goes beyond a mere negotiation of boundaries, between accounts of health and disease (on the one hand) and impairment and disability (on the other); and various aspects of that relationship will be discussed.

Normality, Disorder, and Disability

One aspect that has already emerged in the discussion of models of disability is the relationship between *normality, disorder,* and *disability*. The *ICIDH* model suggests one account of this relationship fairly clearly. *Pathologies* (such as diseases, injuries, or congenital abnormalities) are understood as deviations from normality; they can give rise to *impairments,* understood as losses or abnormalities of structure or function. Impairments can cause *disabilities,* understood as "restriction or lack of ability" to perform activities in a normal manner or range; impairments and disabilities can cause *handicaps,* defined as disadvantages preventing or restricting the performance of normal roles. Thus, at each level of the classification, *normality* plays a key definitional role; and indeed, as we saw earlier, at the level of disability the stipulation is not just that individuals perform in the normal range, but also *in the manner* "considered normal for a human being."[113]

There is a clear connection here with some of the theories of health reviewed in the previous chapter — most obviously that of Christopher Boorse, since he defines health and disease in terms of the performance of natural functions, defined as "species-typical contribution[s] to survival and reproduction."[114] As we have seen, social-model authors contest not only the causal claims in the *ICIDH* account and its individualized conception of disability, but also the work done in that account by the concept of normality.[115] While the *ICF* model does not make such strong or explicit

---

112. Oliver, *Understanding Disability*, p. 45.
113. WHO, *ICIDH*.
114. Christopher Boorse, "What a Theory of Mental Health Should Be," *Journal for the Theory of Social Behavior* 6 (1976): 61-84 (p. 63).
115. Silvers, "The Fatal Attraction of Normalizing"; Amundson, "Against Normal Function." It should be noted that Boorse believes his theory of health to be consistent even with

claims about normality, it still relies on some notion of biological normality at various points, and as such is the object of ongoing suspicion from some disability activists and social-model authors.

Ron Amundson has offered a robust critique of the concept of "normal function" deployed by Boorse and others.[116] Amundson does not, of course, deny that biological structures and functions can be statistically *typical* or *atypical* (in other words, common or unusual within the population); however, he attributes to Boorse a stronger claim about biologically normal function, namely "that functions take place in a uniform mode at a relatively uniform performance level by a statistically distinctive portion of the members of a species. These are the normals."[117] They are clearly differentiated from those statistically unusual members with different functional designs, and different levels *or* modes of performance, from the "normal" population. Amundson argues that this concept of biological normality is supported neither by current evolutionary theory nor by developmental biology. The former recognizes high levels of heritable variability within species and is suspicious of claims about normality. The latter reveals high levels of "plasticity" in the development of individual species members, so that it is misleading to think of development as following a "blueprint": biological goals can be met by highly atypical forms of structural and functional development. According to Amundson, this matters for the theory of health and for disability, because once the concept of normality becomes reified, it all too easily takes on an ideological function in which it underpins questionable normative judgments about quality of life, the normalization of people with impairments, and so on.[118] Boorse's response is (as one might expect) dismissive of Amundson's argument, but it does not really address in any detail his critique of the concept of normality — in particular, the important distinction between species-typicality (which Amundson accepts) and species-normality (which he rejects).[119]

---

the social model, since the latter is best understood not as a scientific claim "that normal human functional ability does not exist," but as an ethical injunction "to redesign the human environment, at any cost, to make it irrelevant" — though he also remarks that he is "not persuaded that social-model theorists have yet said anything either true or useful": Christopher Boorse, "Disability and Medical Theory," in Ralston and Ho, eds., *Philosophical Reflections on Disability*, pp. 55-88 (p. 77).

116. Amundson, "Against Normal Function."
117. Amundson, "Against Normal Function," p. 36.
118. Amundson, "Against Normal Function," pp. 45-51.
119. Boorse, "Disability and Medical Theory," pp. 76-77.

One interesting question is whether other accounts of health and natural function surveyed in the previous chapter are as vulnerable as Boorse's to Amundson's critique, if Amundson is right about "normal function." In particular, would Jerome Wakefield's model of disorder as "harmful dysfunction" be undermined by Amundson's argument?[120]

Some aspects of Wakefield's account might be vulnerable to Amundson's critique, because Wakefield tends to assume (in common with sociobiological and evolutionary-psychological versions of evolutionary theory) that the business of evolutionary studies is to identify human universals that are explicable as evolutionary adaptations. This search for universal adaptations could be understood as a version of the "functional determinism" that Amundson challenges.[121] On the other hand, it is not species-typicality but evolutionary adaptiveness that does the conceptual work in Wakefield's definition of "dysfunction." Furthermore, as we saw in chapter 1, he is quite ready to agree with some of his critics that evolutionarily designed functions often span a wide range, so that the boundaries between normal function and dysfunction are not clear-cut.[122] It is not a large step from there to a version of his model in which diverse forms and functions could be adaptive and could therefore count as "natural functions." Amundson's critique would, however, indicate that a good measure of caution and humility were in order in identifying putative "natural functions" — particularly since the recognition of evolutionary adaptiveness is often a matter of conjecture more than hard evidence. It might also be necessary to incorporate a criterion of developmental integration alongside evolutionary adaptiveness into the concept of natural function, to take account of Amundson's argument that in some cases evolution has selected for developmental plasticity rather than one particular form or structure. That is to say, a physical or mental mechanism could be considered a natural function if it had proved adaptive in the species' evolutionary history, or if a developmental pathway in an individual had resulted in its contribution to that individual's functioning, or both. So to use one of Amundson's examples, sign

---

120. Jerome C. Wakefield, "The Concept of Mental Disorder: On the Boundary between Biological Facts and Social Values," *American Psychologist* 47.3 (1992): 373-388.

121. For Amundson's definition of "functional determinism" — essentially the concept of normal function that he attributes to Boorse — see "Against Normal Function," p. 35.

122. Scott O. Lilienfeld and Lori Marino, "Mental Disorder as a Roschian Concept: A Critique of Wakefield's 'Harmful Dysfunction' Analysis," *Journal of Abnormal Psychology* 104 (1995): 411-420 (p. 414); Jerome C. Wakefield, "Evolutionary versus Prototype Analyses of the Concept of Disorder," *Journal of Abnormal Psychology* 108.3 (1999): 374-399 (p. 382).

and spoken languages would be equally identifiable as natural functions, since there is evidence that pathways of neural development in congenitally deaf individuals result in the involvement of the same neural structures and mechanisms in signing as are used for spoken language in hearing individuals.[123] Therefore a stroke that impaired either a deaf or a hearing individual's use of language (signed or spoken respectively) would be recognizable as a disorder, whereas congenital deafness should instead be understood as an aspect of diversity.

The Limits of "Health" and "Disability"

A second set of questions concerns the scope and limits of the concepts of health and disability. As we saw in the last chapter, theories of health vary widely in the range of human goods that they encompass, from the highly specific and limited to the all-embracing; but how narrow or broad a range of human life should be included in our conception of disability? And should "disability" and "health" be understood as referring to the same, different, or overlapping domains of human experience and activity?

One answer is suggested by some versions of the social model, illustrated by the remark of Michael Oliver quoted earlier: "Most illnesses are treatable and even curable by medical interventions; most impairments are not curable; and all disability can be eradicated by changes to the way we organize society."[124] The view implicit in this remark seems to be that the domain of health and disease is specific, limited, and clearly differentiated from the domains of impairment and disability. This suggests, incidentally, that although biomedical models of health such as Boorse's are regarded with suspicion by social-model theorists,[125] an account such as Boorse's could in fact fit such versions of the social model quite neatly.[126] Boorse's theory is an attempt to define health and disease in specific, limited ways, drawing a clear boundary around the domain of health and making it possible to differentiate that domain from other domains of human life. According to the social model, some of those other domains

---

123. Amundson, "Against Normal Function," pp. 42-43.
124. Oliver, *Understanding Disability*, p. 44.
125. E.g., Amundson, "Against Normal Function."
126. As Boorse himself suggests: "Disability and Medical Theory," p. 77. See further David Wasserman, "Philosophical Issues in the Definition and Social Response to Disability," in Albrecht et al., eds., *Handbook of Disability Studies*, pp. 219-251 (pp. 222-225).

(such as social relationships and political structures) are exactly where disability, properly understood, is located. However, it was argued in the last chapter that Boorse's way of drawing clear boundaries around the domain of health is itself problematic. Furthermore, as we have seen in the present chapter, it is increasingly acknowledged in the social model literature (by various authors including Oliver himself) that the domains of impairment and disability cannot be sharply separated off from one another or from the domains of health and disease.

A very different answer to our question is suggested by Nordenfelt, who places the concept of *ability* at the heart of his theory of health, and therefore takes accounts of health and disability to be concerned with what is essentially the same territory. He also attributes this same assumption to the authors of *ICF*, who, he suggests, have "committed themselves to a strong holistic tradition in the philosophy of health. This can be seen as a continuation of the commitment made by the WHO itself when it proposed its famous definition of health in 1946."[127] It is clear, though, that the authors of *ICF* do not subscribe to the all-embracing concept of health expressed in the WHO definition, since they clearly differentiate between well-being and health. Well-being is defined in *ICF* as "a general term encompassing the total universe of human life domains," whereas health is understood as "a *subset* of domains that make up the total universe of human life."[128]

Nordenfelt's assumption — that the domains of health and disability are more or less the same — leads him to criticize the inclusion of environmental barriers to activity and participation in *ICF*, since he concludes that these "[lie] partly outside the proper domain of health."[129] However, the *ICF* understanding of disability includes not only health but also "health-related domains," namely "those areas of functioning that, while they have a strong relationship to a health condition, are not likely to be the primary responsibility of the health system, but rather of other systems contributing to overall well-being."[130] Thus the inclusion of areas such as education, employment, transport, and the built environment in *ICF* is not a mistaken view that these are aspects of health, as Nordenfelt appears to think. Rather, it follows from the *ICF* authors' deliberate adoption of a

---

127. Nordenfelt, "On Health, Ability, and Activity," pp. 1461-1462. For the WHO definition of health, see above, pp. 2-6.
128. WHO, *ICF*, p. 188 (emphasis added).
129. Nordenfelt, "On Health, Ability, and Activity," p. 1462.
130. WHO, *ICF*, p. 189.

model in which social factors such as these can interact in varied and complex ways with medical factors to affect functioning and disability.

In *ICF*, this broadening-out of disability beyond the domain of health is limited: "[t]he classification remains in the broad context of health and does not cover circumstances that are not health-related, such as those brought about by socioeconomic factors."[131] As such it presumably still invites criticism from a social-model perspective for giving an over-medicalized and insufficiently socialized account of disability. However, it does raise the possibility that *health* should be understood as a limited area of the human good (as argued in the last chapter) but *disability* should refer to a broader range of human activity and experience, albeit one that overlaps with health. In this view, not all impairments would be, or result from, disorders or maladies — though some would.[132] Post-polio paralysis, for example, clearly results from a disease; but it is less obvious that (for example) congenital profound deafness or the cognitive impairments associated with Down syndrome should be identified as disorders.[133]

Health, Disability, and Quality of Life

The prominence of judgments and arguments about quality of life in many areas of health care practice and bioethics is well known. As noted in the Introduction, QoL judgments play a part in arguments and decisions about *(inter alia)* pre-implantation genetic diagnosis, prenatal diagnosis and abortion, neonatal care, genetic enhancement, resource allocation, end-of-life care and withdrawal of life-prolonging treatment, and assisted dying. There I remarked that concepts of health-related QoL (HRQoL) will depend, explicitly or implicitly, on concepts of health. However, the QoL concepts and claims deployed in the bioethical literature, health care practice, and law are frequently criticized by disability activists and philosophers of disability for incorporating prejudicial and poorly-grounded assumptions about disabled people's QoL. A few examples follow.

1. As noted earlier in this chapter, Pfeiffer observes that in some standard measures of HRQoL, wheelchair use (however highly mobile) scores

---

131. WHO, *ICF*, p. 8.
132. And of course this view does not presuppose a simple or one-way causal relationship between impairments and disability.
133. On the latter, see Amos Yong, *Theology and Down Syndrome: Reimagining Disability in Late Modernity* (Waco: Baylor University Press, 2007), pp. 107-116.

lower than walking (however limited).[134] This would appear to reflect a judgment that lack of ability to walk, in and of itself, lowers a person's QoL — a judgment emphatically not borne out by wheelchair users' assessments of their own QoL.[135]

2. Measures such as the quality-adjusted life year (QALY) and disability-adjusted life year (DALY) are increasingly used at a population or healthcare-system level to assess the impact of healthcare interventions or to inform resource allocation decisions. The QALY is a composite measure in which length of life is multiplied by a factor between 0 (representing death) and 1 (perfect health).[136] It is often used to assess both the effectiveness and the cost-effectiveness of health care interventions in terms of increased life expectancy and/or increased QoL. The DALY is also a composite measure, assessing the "burden" of disease by multiplying length of life by a factor from 0 (perfect health) to 1 (death).[137] According to Ian Basnett, both measures can discriminate against disabled people. The DALY, he argues, does not adequately represent the effects of varied social contexts, and privileges professionals' judgments of the burden of disability; insofar as QALY calculations use indices that incorporate questionable judgments about disabled people's QoL, they may systematically discount the benefit and cost-effectiveness of health care interventions for disabled people compared to the same interventions for non-disabled people.[138]

3. Abortion law in the UK currently sets the upper limit for termination of pregnancy at 24 weeks' gestation, unless certain conditions are met, in which case a pregnancy may be terminated at any point up to full term. One of the exceptions to the 24-week limit is "that there is a substantial risk that if the child were born it would suffer from such physical or mental abnormalities as to be seriously handicapped."[139] This legal provision seems to depend on a judgment that serious physical or mental impair-

---

134. Pfeiffer, "The ICIDH and the Need for Its Revision," p. 512.

135. Amundson, "Quality of Life, Disability, and Hedonic Psychology," pp. 374-375; Catriona MacKenzie and Jackie Leach Scully, "Moral Imagination, Disability, and Embodiment," *Journal of Applied Philosophy* 24.4 (2007): 335-351 (esp. pp. 344-345).

136. See, e.g., Alan Williams, "Economics of Coronary Artery Bypass Grafting," *British Medical Journal* 291 (1985): 326-329, and Andrew Edgar, Sam Salek, Darren Shickle, and David Cohen, *The Ethical QALY: Ethical Issues in Healthcare Resource Allocations* (Haslemere: Euromed Communications, 1998). Some versions of the QALY also allow for negative multipliers representing health states worse than death.

137. Basnett, "Health Care Professionals and Their Attitudes," p. 456.

138. Basnett, "Health Care Professionals and Their Attitudes," p. 457.

139. Human Fertilization and Embryology Act, 1990, s. 37 (1) (d).

ments can reasonably be regarded as grave harms — grave enough that very late abortions are justifiable as a means of preventing the birth of children with such impairments. Likewise, it is frequently assumed in the bioethical literature that life with an impairment is of lower quality than without, and this assumption informs arguments about a range of issues from abortion to assisted dying. To give just one example, John Harris argues that those who make use of prenatal diagnosis and abortion to prevent the birth of "handicapped" children are trying to avoid "the wrong of bringing needless suffering into the world."[140] Many disability activists and academics have criticized prenatal diagnosis and abortion as currently practiced on the grounds that it embodies a eugenic agenda and reinforces discriminatory attitudes to disabled people.[141]

As noted earlier, many of these judgments about QoL are challenged by disability activists and authors on the grounds that disabled people's assessment of their own QoL is typically close to non-disabled people's self-assessment, and much higher than non-disabled people's and health care professionals' assessments of disabled people's QoL.[142] Catriona MacKenzie and Jackie Leach Scully argue that this discrepancy arises from the difficulty of imaginatively inhabiting the embodied life of another person, if that life is significantly different from our own; and Ron Amundson cites evidence from hedonic psychology to suggest that disabled people's first-person estimates of their own QoL are likely to be much more reliable than others' third-person estimates of disabled people's QoL.[143] This is not necessarily to dismiss all use of QoL judgments and arguments in health care and bioethics. It does, however, call for a good measure of epistemological humility in making such judgments, a particular wariness of making those judgments about other lives whose conditions are very different from one's own, a willingness to take seriously others' testimony about their own QoL, and an alertness to the ways in which QoL judgments and arguments might enshrine unexamined prejudices about disability.

---

140. John Harris, *Clones, Genes, and Immortality: Ethics and the Genetic Revolution* (Oxford: Oxford University Press, 1998), p. 88.

141. See Shakespeare, *Disability Rights and Wrongs*, pp. 85-102, though Shakespeare does not endorse this critique.

142. This phenomenon is often referred to as the "disability paradox": see, e.g., Gary L. Albrecht and Patrick J. Devlieger, "The Disability Paradox: High Quality of Life against All Odds," *Social Science and Medicine* 48.8 (1999): 977-988.

143. MacKenzie and Scully, "Moral Imagination, Disability, and Embodiment"; Amundson, "Quality of Life, Disability, and Hedonic Psychology."

Illness, Disability, and the Body: Phenomenological Perspectives

In the last chapter we encountered S. Kay Toombs's phenomenological analysis of chronic illness.[144] Toombs's starting point, it will be recalled, is her experience of chronic illness and of a fundamental failure of communication between her physician and herself: it sometimes seemed that they were inhabiting different worlds with different horizons of meaning. Toombs's attempt to understand this experience by way of a phenomenological exploration of illness and the body brings her into contact with those, such as Miho Iwakuma, who have used the phenomenology of Merleau-Ponty (in particular) to explore the experience of impaired embodiment.[145] Iwakuma conceptualizes the experience of impairments in terms of the *"Umwelt"* ("environment") of the people who acquire and live with those impairments. For example, she relates how people come to "embody" assistive devices such as canes and wheelchairs, so that such a device is no longer merely an object separate from the body, but absorbed into its user's *Umwelt*.[146] Thus, wheelchair users develop a sense of the boundaries of their chairs and the space through which they can pass; if someone's chair is touched it can feel as though his or her body has been touched; and Iwakuma describes how members of a wheelchair basketball team "seemed fetishistic and particular about the chairs," customizing and adjusting them "just as one combs hair in a specific way."[147] The process of coming to live with a newly acquired impairment, including the "embodiment" of assistive devices, can be understood phenomenologically as a process of reorganizing one's *Umwelt* according to one's new condition.

As Gareth Williams observes, such phenomenological perspectives fit well with narrative approaches to the sociology of chronic illness and impairment, remarking that "[s]ome (but by no means all) of this [narrative] work can shed a bright light on 'human' conditions."[148] Interestingly for

---

144. S. Kay Toombs, *The Meaning of Illness: A Phenomenological Account of the Different Perspectives of Physician and Patient* (Dordrecht: Kluwer Academic, 1992).

145. Miho Iwakuma, "The Body as Embodiment: An Investigation of the Body by Merleau-Ponty," in Corker and Shakespeare, eds., *Disability/Postmodernity*, pp. 76-87.

146. Iwakuma, "The Body as Embodiment," pp. 78-79. She borrows the term *Umwelt* from the biologist Jakob von Uexküll, "A Stroll Through the Worlds of Animals and Men," in *Instinctive Behavior: The Development of a Modern Concept*, ed. and trans. Claire H. Schiller (New York: International Universities Press, 1957), pp. 5-80.

147. Iwakuma, "The Body as Embodiment," pp. 79-80.

148. Williams, "Theorizing Disability," pp. 131-134 (p. 132).

the account to be developed in later chapters, he also observes that such narratives can express quasi-religious or spiritual concerns, and acknowledges the place of theological traditions in providing a language for exploring those concerns.

However, he cautions that phenomenological and narrative approaches can all too easily become individualist and solipsistic: the story of an individual journey of struggle and self-transformation takes center stage, and the context of social relationships and political-economic structures, to which social-model authors have insistently drawn attention, becomes effaced.[149] A further cautionary note is sounded by Mackenzie and Scully in their treatment of moral imagination and QoL judgments, discussed earlier.[150] Whereas Toombs's phenomenological essay closes with proposals for the renewal of professional-patient relationships, whereby health professionals draw on their own embodied experience to enter empathetically into their patients' lived experience of illness, MacKenzie and Scully emphasize the limits of such imaginative exercises. Drawing on phenomenology and on cognitive-science theories of embodied cognition, they argue that our mental activity, including acts of the imagination, is always embodied,[151] which severely limits our capacity to imagine ourselves

---

149. Williams, "Theorizing Disability," pp. 132-133. He makes a similar critical remark about religious and theological perspectives: "While Christian theology and the learning of other world religions can certainly provide rich languages for exploring questions of ultimate concern . . . they can also — if we are not very careful — reduce the individual to a body and limit the experience of illness and disability to a personal quest for meaning and truth" (p. 132). In later chapters it will be necessary to explore the extent to which this charge against Christian theology is justified, and what resources a Christian theological tradition might have for resisting such dangers.

150. MacKenzie and Scully, "Moral Imagination, Disability, and Embodiment."

151. They bracket out the possibility of disembodied consciousness: "Leaving aside the possibility, which we do not examine further here, of totally disembodied states of consciousness, it is the case that all human beings have/are physical bodies" (p. 342). This raises the interesting question whether a Christian theological anthropology is committed to an understanding of human being as body and soul, which would require a theological account to address these questions in a different way from MacKenzie and Scully. Not necessarily: elsewhere I have suggested that the language of the soul (certainly post-Descartes) is not the most helpful for giving a Christian account of human being, and have favored something like Nancey Murphy's "non-reductive physicalism": Neil Messer, *Selfish Genes and Christian Ethics: Theological and Ethical Reflections on Evolutionary Biology* (London: SCM, 2007), pp. 153-154. But even if we do follow the long Christian tradition of soul-talk, then on a Thomist account (which is far more representative of that tradition than Cartesian dualism), in this life soul and body are closely integrated, and the soul's cognitions are not possible without

being someone whose embodied experience is very different from our own. We are more likely to imagine *ourselves* in the other person's situation, in which case we will project our feelings and evaluations onto that person and his or her situation. For example, if a non-wheelchair user tries to imagine what it is like to be a wheelchair user, she is likely simply to imagine *herself* in a wheelchair, rather than entering imaginatively into the lived experience of someone for whom the chair has become part of her *Umwelt*. We find it equally difficult to imagine ourselves in future conditions that we do not currently experience — so, for example, people who acquire mobility impairments and become wheelchair users report that when it happens, they do not think or feel anything like what they anticipated thinking and feeling.[152] This clearly has important implications both for people's decision-making about their own future health care (problematizing the use of advance directives, for example) and for situations such as those discussed in earlier sections, in which health care professionals or policy-makers might be making decisions on behalf of other people whose embodied experience is very different from their own.

Capabilities, Health, and Disability

The last chapter included a brief discussion of Iain Law and Heather Widdows's account of health as a capability — a combination of "functionings" or "beings and doings" — drawing on the capabilities approach associated with Amartya Sen and Martha Nussbaum.[153] In her book *Frontiers of Justice*, the latter has offered an extended treatment of disability in a capabilities perspective.[154]

Nussbaum's book is a work of political philosophy, taking up three ar-

---

sensory perceptions mediated by the body; while the soul's post-mortem existence and cognition apart from the body are possible by divine grace, they do not represent the ultimate fulfillment of human life: see further below, pp. 141-150.

152. MacKenzie and Scully, "Moral Imagination, Disability, and Embodiment," p. 344.

153. Iain Law and Heather Widdows, "Conceptualizing Health: Insights from the Capability Approach," *Health Care Analysis* 16 (2008): 303-314.

154. Martha C. Nussbaum, *Frontiers of Justice: Disability, Nationality, Species Membership* (Cambridge, MA: Harvard University Press, 2006), chs. 2, 3. In her account, Nussbaum uses the *ICIDH* terminology of impairments, disabilities, and handicaps; in the following discussion I shall not adopt her usage, but stick to the ways in which I have generally used "impairment" and "disability" in this chapter.

eas of unfinished business from the social-contract theory of John Rawls and testing the limits of contractarian political theory in relation to them. One of these areas is justice for disabled people. Nussbaum argues that Rawls cannot give an adequate account of the just treatment of disabled citizens, for various reasons. One is that his well-known thought-experiment of the Original Position assumes that the parties to the social contract have "native endowments such as strength and intelligence" lying "within the normal range,"[155] and are able to be "normal and fully cooperating members of society over a complete life."[156] Another is that his Kantian account of the value of human persons entails a sharp distinction between the deterministic realm of nature and the realm of human freedom: "[i]t is in virtue of our capacity for moral rationality, and that alone, that we rise above [the realm of nature] and exist, as well, in a realm of ends."[157] Some people with physical impairments are in danger of being excluded by the first of these features of Rawls's theory, but those with severe mental impairments face the double jeopardy of exclusion both from the status of parties to the contract in the Original Position and from this Kantian conception of the value of persons.

Nussbaum argues that a capabilities approach informed by a neo-Aristotelian conception of human flourishing is better able to handle issues of justice for disabled people. In contrast to Sen (who is reluctant to draw up a list) she argues that at least a working list of core or basic human capabilities is needed. She proposes ten: life; bodily health; bodily integrity; use of the senses, imagination, and thought; emotions; practical reason; affiliation (including the social bases of self-respect and non-humiliation); relationship with other species; play; control over one's environment, both political and material.[158] She argues that this list holds good across cultures and for all citizens. The meeting of at least a threshold level of each capability is a requirement for "a life that is worthy of the

---

155. John Rawls, *A Theory of Justice* (Cambridge, MA: Harvard University Press, 1971), p. 25.

156. John Rawls, *Political Liberalism*, expanded ed. (New York: Columbia University Press, 2005), pp. 20, 21. This assumption is derived from the basic idea of the contractarian tradition, that parties leave the "state of nature" to contract together *in order to gain a mutual advantage*.

157. Nussbaum, *Frontiers of Justice*, p. 131.

158. E.g., Martha Nussbaum, "Women's Capabilities and Social Justice," *Journal of Human Development* 1.2 (2000): 219-229; see also Martha Nussbaum, *Women and Human Development* (Cambridge: Cambridge University Press, 2000).

dignity of a human being"[159] — those prevented from achieving the threshold level of any of these capabilities have been treated unjustly, their dignity affronted. Furthermore, the list is irreducibly plural, so increasing the level of one capability does not compensate for a deficit in another.[160]

Nussbaum shows convincingly that her capabilities approach gives a more satisfactory account than social-contract theories of the requirements of a just society in respect of its disabled members, in areas such as the provision of care, education, and political participation. More generally, in the context of the present project, the Aristotelian roots of her account make possible a richer conception of human flourishing than some of the alternatives on offer. Nonetheless (leaving aside any more general critical discussion of her approach as a piece of political philosophy), in the context of this chapter's reflections on disability, Nussbaum's account raises difficult issues. In particular, one area of difficulty relates to her insistence that the same list of basic capabilities and the same thresholds must apply to all citizens, with or without impairments. The importance of this stipulation is easy to understand, particularly in the political-philosophy context to which her book is addressed. Conceding that some capabilities do not apply to people with impairments, or that lower thresholds apply to them, makes it all too easy for a society to avoid its obligations to provide the support that will enable them to achieve a high level of functioning, rationalizing its lack of support on the grounds that such people have natural and inevitable limitations. For example, "many problems of children with Down syndrome that had been taken to be unalterable cognitive limitations are actually treatable bodily limitations," whose proper treatment at the right time can dramatically enhance those children's cognitive development.[161]

Important though this is, it does have some costs and dangers, illustrated by Nussbaum's discussion of a young woman with severe cognitive impairments who is most unlikely to achieve the threshold level of some of the basic capabilities:

> It now would appear that the view that emphasizes the species norm must choose: either we say that Sesha has a different form of life alto-

---

159. Nussbaum, *Frontiers of Justice*, p. 180.
160. Nussbaum, *Frontiers of Justice*, pp. 166-167.
161. Nussbaum, *Frontiers of Justice*, pp. 186-190 (p. 189).

gether, or we say that she will never be able to have a flourishing human life, despite our best efforts.[162]

In relation to Sesha, Nussbaum answers her own question by affirming that Sesha's life is a recognizably human life, because "at least some of the most important human capabilities are manifest in it, and these capabilities link her to the human community rather than some other"[163] — in which case, she is presumably committed to saying that Sesha is a human being who in some respects will never flourish. The danger here is that her account, or the uses to which it is put, will drift towards a conception of normality of the kind criticized by Ron Amundson, Anita Silvers, and others.[164] In relation to more extreme impairments, however, Nussbaum chooses the other alternative: "Only sentiment leads us to call the person in a persistent vegetative condition, or an anencephalic child, human."[165] In this respect her list of basic capabilities functions in the way in which the concept of "person" often functions in contemporary bioethics, to mark out a boundary that excludes some offspring of human parents from the status of "human" — albeit a more hospitably drawn boundary than in some accounts. Although this is a common move in bioethics, it is one that many critics find troubling, and one to which the theological account to be developed in later chapters is likely to take exception.[166]

Teleology, Health, and Disability

Toward the end of the last chapter I discussed the prospects for a teleological account of health and disease, with reference to Christopher Megone's presentation of Aristotelian teleology. Megone's Aristotelian theory of health includes the following key features:[167]

---

162. Nussbaum, *Frontiers of Justice*, p. 187. The young woman in question is Sesha Kittay, daughter of the philosopher Eva Feder Kittay and her husband Jeffrey: see Nussbaum, *Frontiers of Justice*, p. 96, citing Eva Feder Kittay, *Love's Labor: Essays on Women, Equality, and Dependency* (New York: Routledge, 1999).
163. Nussbaum, *Frontiers of Justice*, pp. 187-188.
164. Amundson, "Against Normal Function"; Silvers, "The Fatal Attraction of Normalizing."
165. Nussbaum, *Frontiers of Justice*, p. 187.
166. See Neil Messer, *Respecting Life: Theology and Bioethics* (London: SCM, 2011), pp. 114-118, for a theological critique of such uses of "personhood."
167. Christopher Megone, "Aristotle's Function Argument and the Concept of Mental

1. Humans can be understood as members of a natural kind, which undergoes certain characteristic changes open to teleological explanation — that is, explanation in terms of their contribution to goals that are recognizable as good. Put another way, these are the changes characteristic of a *good* member of the species.
2. Goals that are recognized as good are those that contribute to the survival and persistence of the species.
3. Natural functions can be understood as contributions to the cycle of changes exhibited in a good human life — that is, those changes that contribute to the survival and persistence of the species.
4. For human beings, the functions that contribute to a good life of this kind must have to do with what is distinctive about the human species, and that turns out to be *rationality* (broadly defined).
5. Diseases and illnesses can be understood as losses of function that inhibit the capacity for a rational life (broadly understood).

While I raised some important critical questions about Megone's account, I suggested in the last chapter that *some* form of teleological understanding of health in terms of human flourishing will prove important for the theological account to be developed later in the book. However, a disability perspective could be expected to raise some critical questions for claims about human flourishing and teleology, some of which have already been hinted at in earlier sections.

Garret Merriam is one recent author who has explored these connections.[168] Acknowledging what presumably hardly anyone would deny — that Aristotle's own writings make some very troubling claims about disabled persons, as they do about women and slaves — Merriam argues that Aristotelian virtue ethics can nonetheless be "rehabilitated," as he puts it, to give an account of human flourishing that is illuminating in relation to disability.

One aspect that he thinks is in need of rehabilitation relates to the first point above: the understanding of humans as a natural kind. His concerns in this regard are related to those of Amundson, discussed ear-

---

Illness," *Philosophy, Psychiatry, and Psychology* 5.3 (1998): 187-201; Christopher Megone, "Mental Illness, Human Function, and Values," *Philosophy, Psychiatry, and Psychology* 7.1 (2000): 45-65.

168. Garret Merriam, "Rehabilitating Aristotle: A Virtue Ethics Approach to Disability and Human Flourishing," in Ralston and Ho, eds., *Philosophical Reflections on Disability*, pp. 133-151.

lier.[169] Merriam attributes to Aristotle a kind of "species essentialism," which holds that there is "an eternal fundamental nature that constitutes the species . . . [a] single, absolute, metaphysical, archetypal human being by which the rest of us are to be measured."[170] Regardless of whether this fairly represents Aristotle's view (and in an endnote he goes some way to acknowledging that it might not),[171] it is widely held that after Darwin such a view is untenable. However, Merriam attributes a different form of "species-standard metric" to post-Darwinian accounts such as Christopher Boorse's biostatistical theory of health. Arguing that neither Aristotle's account of flourishing nor Boorse's theory of health can distinguish properly between disease and disability, he recommends instead an account of flourishing informed by the Stoics. In this account the proper question to ask is not, "how does this individual compare to a species-norm in terms of the capacities necessary for flourishing?" but *"given the individual circumstances of this person's life, are they living well, or living poorly?"*[172] That question is to be answered by the application of *phronēsis*, or practical wisdom.

One question to be asked about this is, what will give *phronēsis* its moorings, so to say?[173] While Merriam is correct to suggest that virtue eth-

169. Amundson, "Against Normal Function."
170. Merriam, "Rehabilitating Aristotle," p. 134.
171. Merriam, "Rehabilitating Aristotle," p. 148 n. 10.
172. Merriam, "Rehabilitating Aristotle," p. 135, emphasis original.

173. Martha Nussbaum's discussion of disability and social justice suggests another hostage to fortune given by Merriam's reliance on the Stoics to correct Aristotle. She points out that the Stoics, in contrast to Platonic and Aristotelian theories that "saw considerable continuity between human capacities and the capacities of other animals," insisted on making the sharpest of distinctions between humans and "the beasts"; in Nussbaum's words, "[p]ejorative remarks about animals frequently substitute for argument in [the Stoics'] accounts of human nature and human dignity" (Nussbaum, *Frontiers of Justice*, pp. 130-131). As Nussbaum makes clear, the long tradition of reflection on personhood following from the Stoics — including, notably, the thought of Immanuel Kant — has tended to privilege reason as what makes humans valuable. Thus, it is clear how a Stoic conception of the good life could make it easy for Merriam to recognize the life of an intelligent and high-achieving woman such as Helen Keller as one lived "very, very well" ("Rehabilitating Aristotle," p. 138); but his account might make it harder to acknowledge the same of someone with a severe cognitive impairment or someone whose life was marked by a much higher degree of dependency. Quite apart from the disparagement of non-human animal life implicit in such a view (for a range of critical perspectives on which, see Celia Deane-Drummond and David Clough [eds.], *Creaturely Theology: On God, Humans, and Other Animals* [London: SCM, 2009]), there is in short a worry that Merriam's conception of a good human life has an

ics does not depend on a single guiding moral principle or calculus, there must be *something* that gives us criteria for making a judgment in answer to the question. To take one of Merriam's paradigm cases, Helen Keller, there must be something that enables us to recognize (for example) gaining a degree and cofounding the American Civil Liberties Union as expressions of a life lived well rather than badly.[174] *Phronēsis*, in other words, is not a free-floating virtue that operates in isolation from others — as is made explicit in some important accounts influenced by Aristotle. For Thomas Aquinas, for example, *prudentia* (his Latin equivalent of *phronēsis*) is closely linked to the other cardinal virtues of courage, temperance, and justice, being concerned with their application to specific situations. The exercise of *prudentia*, in Jean Porter's words, "determines what amounts to a substantive theory of the human good, at least as it applies to this individual in his particular setting."[175]

Even if *phronēsis* does need some kind of anchoring point to enable judgments about whether a life is lived well or poorly, however, is the concept of a human natural kind the right place to fasten an anchor? Both Amundson and Merriam have in effect raised two critical questions about this use of a natural-kind concept. One is whether it is sustainable in the light of modern biology; the other is whether it inevitably privileges statistically typical body structures and functions, illegitimately labeling those with less typical bodies or modes of functioning as "abnormal" or "defective."

Daniel Sulmasy has defended a notion of natural kinds that he believes can be helpful in understanding and responding to disability. In his account (drawing on Saul Kripke, Baruch Brody, and David Wiggins), the

---

inbuilt bias in favor of the rational and active life and against those forms of human life that seem deficient in these respects. As we shall see in the next chapter, a number of theologians who reflect on disability argue that the lives of those with cognitive impairments call into question precisely such assumptions about what makes a human life a good one.

174. It might be replied that the criterion by which an individual life is judged well or poorly lived is simply how well it measures up to the "vital goals" (to borrow Nordenfelt's phrase) that the individual chooses for him or herself. But apart from the difficulties, noted in the previous chapter, with this kind of subjective and individualist notion of vital goals, to posit this criterion is *already* to state a particular theory of the human good, one in which the human good consists in each individual's freedom to choose his or her vital goals: see Alasdair MacIntyre, *Whose Justice? Which Rationality?* (London: Duckworth, 1988), pp. 335-338.

175. Jean Porter, *The Recovery of Virtue: The Relevance of Aquinas for Christian Ethics* (London: SPCK, 1994), p. 162.

naming of a kind is an act of recognition of a reality that is there in the world: "that there is something about each natural kind that is common to all members of the kind and yet distinct from all other kinds — a 'kindedness,' if you will — that precedes the naming of the kind."[176] He believes that one of the characteristic features of the human kind is its very *variability*. Indeed, he observes that the genetic component of that variability, "over evolutionary time, is vital to the continued flourishing of the kind."[177] He classifies biological differences into three types: *variations*, which are irrelevant to individual and species flourishing; *differentiations* (such as sex and developmental stage), which are necessary for the flourishing of the kind; and "diseases, injuries, and disabilities," which, "of themselves, inhibit the possibilities for the affected individual to flourish as the kind of thing that each is."[178]

In light of the discussion in this chapter, there are two related problems with this account. First, Sulmasy associates disabilities with diseases and injuries, begging the causal question raised by the social model in favor of a medical view in which "[m]ost disabilities result from disease or injury."[179] In itself, this is partly a matter of language, and reflects a usage that is more common in American than British disability studies. He certainly recognizes that many of the harms associated with disability are harms of social discrimination. If he were to say that "most *impairments* result from disease or injury," and reserve the term "disability" to refer to the whole complex interaction between impairments, individuals' responses to their impairments, and their social context, there would be little distance between his account and (for example) *ICF*.

Language, however, is not the only issue to be raised. The second problematic feature of Sulmasy's account is its assumptions about the types of biological difference that inhibit flourishing — assumptions that would be strongly contested by Amundson, among others. For example, Sulmasy claims that even without specialized knowledge, "one can readily recognize" children born with foreshortened limbs both as *members* of the human natural kind and as "defective" members of this kind.[180] Likewise, he identifies

---

176. Daniel P. Sulmasy, "Dignity, Disability, Difference, and Rights," in Ralston and Ho, eds., *Philosophical Reflections on Disability*, pp. 183-198 (p. 187).
177. Sulmasy, "Dignity, Disability, Difference, and Rights," p. 192.
178. Sulmasy, "Dignity, Disability, Difference, and Rights," p. 193.
179. Sulmasy, "Dignity, Disability, Difference, and Rights," p. 188.
180. Sulmasy, "Dignity, Disability, Difference, and Rights," p. 188. It should be emphasized that he argues in a later section that the value or dignity of an individual member de-

congenital cochlear nerve deficiency as a disease that *disables* its bearer "with respect to the ability to use spoken language," and argues that "necessarily, it is by noticing the deviation from the law-like generalizations and typical features and history of members of the kind that one makes the diagnosis of a disease . . . and concludes that [its bearer] is disabled."[181] This exposes a deep disagreement with (for example) Amundson about the sorts of biological variation that inhibit flourishing, and how they should be recognized. Sulmasy depends on the notion of species-normality (albeit in a modest and critical form) that both Merriam and Amundson criticize. "Without at least this much essentialism," he claims, "medicine would not even be conceptually possible."[182]

Sulmasy's claim, that medicine depends on the concepts of a natural human kind and disease as a state that inhibits the flourishing of members of that kind, has a good deal of plausibility — though of course not everyone will agree with it. But does it, as he thinks, depend on a concept of species-normality to distinguish between diseases and other kinds of variation that do not inhibit flourishing? An answer to this question might be suggested by extending his observation that biological variability itself is "one of the law-like generalizations and typical features and history of the human kind."[183] Sulmasy recognizes the evolutionary importance of *genetic* variability, as does Amundson. However, the latter also emphasizes the importance of developmental plasticity as another source of variation, and claims there is abundant evidence that among members of the same species, biological goals can be achieved with comparable success by means of a wide variety of anatomical and physiological designs.[184] Following Jiří Vácha and others, Amundson proposes a concept of "responsiveness" or "individual normality," which compares the individual's actual performance with the performance necessary to him or her, rather than to a statistical norm.[185]

---

pends *only* on "the bare fact of membership in the natural kind . . . it is not being black or white, male or female, able or disabled that grounds arguments for dignity. Rather, all that counts is being human" (p. 191).

181. Sulmasy, "Dignity, Disability, Difference, and Rights," p. 190.
182. Sulmasy, "Dignity, Disability, Difference, and Rights," p. 188.
183. Sulmasy, "Dignity, Disability, Difference, and Rights," p. 192.
184. Amundson, "Against Normal Function," pp. 38-44.
185. Amundson, "Against Normal Function," p. 44. There could of course be many areas of medicine where species-typicality remains a useful diagnostic criterion, and deviation from a statistical norm is a valuable indicator that something is wrong or at risk of going

Perhaps it is possible to articulate a concept of the human natural kind that can address these concerns. We might say that the human kind has certain characteristic [biological] goals, the achievement of which is at least part of what it means for a member of this kind to flourish. A disorder or disease could be understood as an internal state that tends to inhibit the realization of one or more of those goals. But one characteristic feature of this kind, as Sulmasy recognizes, is *variability* — not only genetic, but also in terms of developmental plasticity and the range of ways in which anatomical and physiological systems can be configured to achieve biological goals. We should therefore be very wary of stipulating that more common ways of achieving biological goals (for example, the use of spoken language to achieve the goal of communication) are more conducive to the flourishing of this kind than more unusual ways (for example, signing).[186]

A further area of contention from a disability perspective concerns the goals that might be identified as characteristic of the human kind. For instance, as Megone shows, Aristotle identified the life of reason as characteristic of the human species;[187] but those with experience of cognitive impairments might well protest against the centrality given to reason in this teleological account of human functions. Against this objection it could be responded that Aristotle's notion of rationality is much broader than mere cognitive capacity, encompassing such things as empathy and sociality, so that an individual with Down syndrome (for example) could exhibit to a high degree what Aristotle had in mind. But this is at best a partial answer, and it must be recognized that one important contribution of a disability

---

wrong. If I had blood pressure of 180/110 or a temperature of 40° C (104° F), I would certainly not want my doctor to ignore these facts. I take Amundson's complaint not to be against this kind of diagnostic use of statistical norms, but the use of statistical normality to define what *counts* as flourishing, so that unusual but effective ways of achieving the organism's biological goals are classed as defective, and potentially inappropriate and unwelcome forms of medical intervention and rehabilitation are inflicted on those who exhibit them.

186. It might be objected that some of the variations Amundson has in mind, such as foreshortening or absence of lower limbs, *do* seriously inhibit the achievement of biological goals unless they are compensated for by technological aids such as wheelchairs, and therefore should be counted as disorders or defects, as Sulmasy holds. However, since a characteristic feature of the human kind is to depend on technological manipulations (broadly understood) of our environment to meet even our most basic needs — such as agriculture to meet our need for food — it does not seem very promising to use dependence on technology as a criterion for identifying defective states or variants; see Amundson, "Against Normal Function," p. 45.

187. Megone, "Aristotle's Function Argument," p. 195.

perspective to the present project is to raise critical questions about the distorting effects of prejudice, projection, and ideology on conceptions of human flourishing. Perhaps rationality, however understood, is not so central to the good of the human kind as Aristotle imagined.

Critical Questions for Concepts of Health, Disease, and Illness

On the basis of the foregoing reflections on disability, some critical questions can be put to the discussion of health and disease in the previous chapter. These critical questions will also be important to keep in mind as a theological account of health is elaborated in subsequent chapters.

1. One of the key questions raised by the social model of disability is whether we misrepresent the causes of human failures to flourish. Are we too ready to attribute any lack of flourishing to individual and medical, rather than social, political, or economic, causes? Or, more generally, if flourishing is inhibited by a complex interaction of individual and social causes, are we too ready to notice the individual and ignore the social dimensions?

2. If we do misrepresent the causes of failures in flourishing, does this also lead us to misdirect our efforts to safeguard and promote flourishing? For example, do our societies put resources into medical interventions or rehabilitation that would be better directed at the removal of environmental, social, and political barriers to activity and participation, as social-model authors argue?

3. Do we tend to misrepresent bodily diversity as pathology? In other words, do we have too narrow a view of the range of different ways in which embodied human life can flourish — perhaps motivated by prejudices and aesthetic preferences for the statistically "normal," as Amundson, Silvers, and others argue?

4. A related question concerns who has the authority to determine whether a condition should be considered a disorder or an expression of diversity. More generally, whose view of whether a condition causes distress, dysfunction, or reduced quality of life should be given most weight? This question is raised, for example, by the striking differences between first-person and third-person estimates of many disabled people's QoL, discussed by Amundson, Basnett, and others. A particularly sharp critical perspective on this issue might be offered by the Foucauldian critique, discussed by Shelley Tremain and others, that discourses of disease and impairment function as exercises of knowledge/power with a disciplinary purpose.

5. If we are prone to misrepresent diversity as disorder, is there also a risk that medical and other interventions intended to promote bodily and mental flourishing will inhibit it for some people, by "correcting" differences that are identified as threats to flourishing but which should rather be understood as *aspects* of those individuals' flourishing? One classic example is the controversy over whether profoundly deaf young children should be given cochlear implants, which might give them some measure of hearing but cut them off from the sign-language culture of the Deaf community.[188] Ron Amundson discusses other instances where, he argues, "normalizing" approaches to treatment, rehabilitation, and education have actually inhibited the flourishing of people with (for example) foreshortened limbs, impaired mobility, and autism.[189]

6. The Foucauldian critique mentioned above also raises a more general critical question about evolutionary and other biological accounts of natural function in relation to health and disease, such as Wakefield's "harmful dysfunction" model. Are evolutionary claims about natural function best understood not as objective scientific insights, but legitimations of an ideological commitment to normalizing of the kind criticized by Amundson, Silvers, and others? Or more modestly, even if one does not attempt a full-scale poststructuralist deconstruction of all scientific truth-claims,[190] is there a danger that theorists' particular social locations or ideological commitments will bias or distort evolutionary claims about natural function and the notions of health and disease that they inform? In related areas of human evolutionary theorizing, it has frequently been argued that the ideological commitments and social locations of in-

---

188. See Robert Sparrow, "Defending Deaf Culture: The Case of Cochlear Implants," *Journal of Political Philosophy* 13.2 (2005): 135-152. As Sparrow makes clear, however, this example is complicated by the insistence of many members of the Deaf community that it should not be considered a community of people with a particular class of impairments, but a cultural and linguistic minority (esp. pp. 139-140). See further Mairian Corker, "Deafness/Disability — Problematizing Notions of Identity, Culture, and Structure," in *Disability, Culture, and Identity,* ed. Sheila Riddell and Nick Watson (London: Pearson, 2003), ch. 6.

189. Amundson, "Against Normal Function," pp. 48-51.

190. For a helpful discussion of how scientific medicine might be located in a postmodern context of a range of "body knowledges" (focusing on a molecular rather than evolutionary paradigm, though the two are of course related), see Jackie Leach Scully, "A Postmodern Disorder: Moral Encounters with Molecular Models of Disability," in Corker and Shakespeare, *Disability/Postmodernity,* pp. 48-61; and for a case for a critical realist rather than radically constructionist approach to disability, Shakespeare, *Disability Rights and Wrongs,* pp. 54-55.

vestigators have biased the research questions they asked and therefore the answers they received.[191] At any rate, this suggests that evolutionary claims about natural function are best made in a cautious and chastened manner, particularly since evolutionary claims about human nature are often matters of conjecture, drastically under-determined by hard evidence — as Wakefield seems to acknowledge.[192]

7. Since I have suggested at various points that a teleological view of health will be a promising direction to follow in developing a theological account, a similar critical question must be raised about teleological accounts. If we make claims about humans as a natural kind with characteristic goods and ends, we must at any rate be very cautious that our claims about the proper ends and goals of embodied human life are not simply projections of the contingent experience of the majority, or of those whose perspectives happen to dominate the discussion.

With these questions, and the agenda identified at the end of the last chapter, in mind, we are now ready to begin identifying the resources and elements that will be needed for a theological account of health, disease, and illness that can address this agenda and these questions. That will be the task of the next chapter.

---

191. See further Messer, *Selfish Genes and Christian Ethics,* pp. 118-119, 177-178, and references therein.

192. Wakefield, "The Concept of Mental Disorder," pp. 382-383. An acknowledgement of the provisional and conjectural nature of evolutionary claims about natural function seems to be reinforced by his subsequent proposal that natural function should be understood as a "black box essentialist" concept: Jerome C. Wakefield, "Aristotle as Sociobiologist: The 'Function of a Human Being' Argument, Black Box Essentialism, and the Concept of Mental Disorder," *Philosophy, Psychiatry, and Psychology* 7.1 (2000): 17-44.

CHAPTER 3

Theological Resources for Understanding Health and Disease

Introduction

Previous chapters have offered critical surveys of academic debates in the philosophy of medicine, disability studies, and related fields. The accounts discussed in those chapters drew in their turn on other academic disciplines such as the biomedical sciences: several of the accounts of health and disease surveyed in chapter 1, for example, drew more or less extensively on evolutionary biology. Since the project of this book is to develop a *theological* account of health, disease, and illness, questions arise about what sources should be used to develop that account, and how it should engage with these other academic discourses. More specifically, in developing a theological account of health, how should the distinctive sources of Christian faith and practice — the Scriptures and the Church's ongoing tradition of reflection on them — interact with insights and arguments from other academic disciplines such as biology or philosophy?

This issue could be framed in a rough-and-ready way as a question about the relationship of sources "internal" and "external" to Christian faith and practice in shaping an understanding of health, disease, and illness. A range of possibilities can be identified.[1] At one extreme would be

---

1. It will be obvious that this typology owes a good deal to Hans W. Frei, *Types of Christian Theology*, ed. George W. Hunsinger and William C. Placher (New Haven: Yale University Press, 1992); see also David Ford, "On Being Theologically Hospitable to Jesus Christ: Hans Frei's Achievement," *Journal of Theological Studies* 46.2 (1995): 532-546. I first developed a version of it in Neil Messer, *Selfish Genes and Christian Ethics: Theological and Ethical Reflections on Evolutionary Biology* (London: SCM, 2007), pp. 49-61, where it is expounded and

the view that our understanding of these concepts must be determined entirely by external influences such as philosophical arguments and scientific data, while at the other would be the insistence that only internal sources such as biblical texts should have any influence whatever on what we understand by health and disease. Neither of these extremes seems very promising for the development of a theological account, so we are left with a series of intermediate possibilities. One is that the account is shaped primarily by external sources such as philosophy and the biosciences, but distinctively Christian sources such as Scripture can supplement or modify what we learn from the former. Alternatively we might proceed by way of what Hans Frei calls "ad hoc" correlations between internal and external sources,[2] neither kind of source having a dominant role in shaping our account. Or again, our account might be shaped primarily by internal sources, insights from external sources being critically appropriated to it.

This last mode of theological engagement is the one adopted in what follows. My aim will be to develop a theological understanding of health, disease, and illness shaped primarily by a reading of the distinctive sources of Christian faith: primarily Scripture, and secondarily Christian traditions of reflection and practice developed in response to the Scriptures. This choice reflects a general commitment to a certain kind of theological approach, characteristic of a broadly Reformed theological tradition, which I do not have space to defend fully here. However, one reason for advocating such a mode of engagement in the present work is that if theology has anything to contribute to an understanding of health, disease, and illness, it is most likely to be able to make that contribution to the extent that it is free to speak in its own characteristic voice, out of its own distinctive sources and approaches. A theologian working in this way should certainly be ready to engage fully and carefully with the kinds of academic discourse I have surveyed in the first two chapters; but he or she should not feel constrained simply to accept the agenda as it is set by those other disciplines. The task of such a theological engagement on the topic of health, disease, and illness is not merely to try to answer the questions as they are framed in (for example) the philosophical literature, but to reframe those questions in the light of theology's distinctive sources and methods. Some in-

---

discussed more fully. The labels "external" and "internal" to describe the different kinds of source are of course not entirely satisfactory, but I hope it is clear enough what is meant by them.

2. Frei, *Types of Christian Theology*, pp. 70-78.

stances of this reframing will emerge in this and the next chapters. One that serves to illustrate the point is that Aristotelian accounts of health raise the question whether it is possible, after Darwin, to think of the human species as a natural kind with its own characteristic ends; but I shall argue that Christians should instead think of humans as *creatures* of a particular kind, using a distinctively theological category that makes explicit reference to a Creator. As I shall try to show, changing the category in this way makes a real difference to the way we think about the human kind and its good.

More generally, it has already become clear in the first two chapters that discussion of health, disease, and illness tables some very deep questions about what it is to be human, how we are to understand human flourishing or the human good, what sense (if any) we can make of suffering and evil, and more besides. In the theological tradition being articulated here, the final authority for answering such questions is to be found in God's self-revelation in Jesus Christ, to which the Scriptures bear witness. A theologian working in the tradition should therefore be unwilling to concede in principle the *autonomy* of the other disciplines with which this book engages in addressing these questions. Once that is acknowledged, however, such a theological approach can be thoroughly hospitable to insights of all kinds from other disciplines and perspectives, provided always that those insights are *critically* appropriated according to theological criteria. Theologians who understand their task as responding faithfully to the one Word of God and Light of the world, Jesus Christ, have no reason to expect that God's word will be spoken and God's light will shine *only* within the "sphere of the Bible and the church," as Karl Barth puts it: they "can and must expect that [God's] voice will also be heard without."[3]

I have remarked that the theological tradition in which I am working can be identified as broadly Reformed, and one of the major resources for the following account is Barth, the most important theologian of that tradition in the past century. A theological approach informed by him will, as I have already indicated, be from first to last a "theology of the Word": a response first and foremost to Jesus Christ, the Word made flesh, and therefore also to the written Word in Scripture, which bears witness to the incarnate Word. It should be no surprise, therefore, that the first two sections of this chapter begin with biblical texts, and there is further bibli-

---

3. Karl Barth, *Church Dogmatics*, 13 vols., ed. and trans. Geoffrey W. Bromiley and Thomas F. Torrance (Edinburgh: T&T Clark, 1956-75), IV.3.1, p. 117. Hereafter *CD*.

cal exposition at many points. However, there are limits to my direct engagement with biblical material. While some key texts are expounded directly, and quotations, citations, and allusions indicate the presence of many more in the background of the theological themes I discuss, I do not offer anything like a full survey of relevant biblical sources. The reader is referred to other recent works for a more complete picture of this aspect than I can give within the constraints of the present work.[4]

While I have located my own account within a Reformed tradition influenced by theologians such as Barth, I also seek to maintain an ecumenical openness to conversations with other theological traditions. Thus, another of my conversation partners in this chapter is Thomas Aquinas. Clearly there are points of quite sharp tension between this theological conversation and the one I conduct with Barth, and those tensions will be discussed later in the chapter. In chapter 4 I shall try to synthesize insights from these two very different theological corpora into one coherent account, an attempt that seems slightly less foolish in the light of a growing body of recent and current work on the relationship between Barth's and Thomas's thought.

The Christian tradition, however, is not only a tradition of belief and understanding but also of practice; indeed, the theological discourse of the tradition can be understood to an important extent as an historically extended critical reflection on the Christian community's practices of worship, common life, and service. Since the 1970s, many theologians have followed in the footsteps of Stanley Hauerwas by emphasizing the importance of the Church's distinctive practices for the formation of its theological and moral vision, and for enabling its members to live by that vision. Therefore alongside the theological thought of Barth and Thomas (neither of whose intellectual work, of course, could exactly be described as *detached* from the context of Christian community and practice), I try to attend in the present chapter to aspects of Christian practice that are particularly significant for the questions being addressed in this book.

It will be evident from what has already been said that the resources for understanding health and disease within the Christian tradition are many and diverse. Of those diverse resources I explore just four in this chapter. My selection has been guided partly by the general considerations of theological approach that have been sketched out in this introduction, and

---

4. For one helpful theological and pastoral engagement with a wider range of biblical texts, see Frederick J. Gaiser, *Healing in the Bible: Theological Insight for Christian Ministry* (Grand Rapids: Baker Academic, 2010).

partly to facilitate a theological engagement with the discussions surveyed in earlier chapters. The chapter begins with a highly selective look at Christian practices of healing and care of the sick, at the biblical texts that underpin these practices, and recent reflection on those texts. Next, having identified Karl Barth as an important conversation partner in the theological tradition within which my work is located, I discuss his own extended account of health and sickness. Third, I explore the teleological understanding of human nature found in the thought of Thomas Aquinas, a theme which takes up (in a very different way) a teleological emphasis implicit in Barth's account, and makes possible a critical engagement with the philosophical discussion of teleology surveyed in chapter 1. Finally (again, highly selectively) I discuss some recent theological writing on disability, thus engaging theologically with the critical perspectives raised by the study of disability in chapter 2. Again in this final section, attention to the practice of the Christian community is one important source of my reflections.[5]

Christian Practices of Healing and Care for the Sick

On the way to Jerusalem Jesus was going through the region between Samaria and Galilee. As he entered a village, ten lepers approached him. Keeping their distance, they called out, saying, "Jesus, Master, have mercy on us!" When he saw them, he said to them, "Go and show yourselves to the priests." And as they went, they were made clean. Then one of them, when he saw that he was healed, turned back, praising God with a loud voice. He prostrated himself at Jesus' feet and thanked him. And he was a Samaritan. Then Jesus asked, "Were not ten made clean? But the other nine, where are they? Was none of them found to return and give praise to God except this foreigner?" Then he said to him, "Get up and go on your way; your faith has made you well." (Luke 17:11-19, NRSV)

Are any among you suffering? They should pray. Are any cheerful? They should sing songs of praise. Are any among you sick? They

---

5. For an excellent contrasting account of health drawing extensively on Patristic sources (particularly Augustine), which nevertheless comes to somewhat similar conclusions to my account, see Almut Caspary, *In Good Health: Philosophical-Theological Analysis of the Concept of Health in Contemporary Medical Ethics* (Stuttgart: Franz Steiner, 2010).

should call for the elders of the church and have them pray over them, anointing them with oil in the name of the Lord. The prayer of faith will save the sick, and the Lord will raise them up; and anyone who has committed sins will be forgiven. Therefore confess your sins to one another, and pray for one another, so that you may be healed. The prayer of the righteous is powerful and effective. Elijah was a human being like us, and he prayed fervently that it might not rain, and for three years and six months it did not rain on the earth. Then he prayed again, and the heaven gave rain and the earth yielded its harvest. (James 5:13-18, NRSV)

The Christian Church has a long and diverse tradition of the practice of healing and the care of the sick. The contemporary exponents of that tradition trace it back to texts like the two just quoted: the Gospel narratives of Jesus' healing ministry, and texts like the one from the letter of James, which reflect the healing practice of early Christian communities.[6] Histor-

---

6. For just two among many recent discussions that take the Gospel narratives in particular as paradigmatic for the contemporary Christian healing ministry, see John Perry, ed., *A Time to Heal: A Report for the House of Bishops on the Healing Ministry* (London: Church House Publishing, 2000), pp. 26-34, and Abigail Rian Evans, *Redeeming Marketplace Medicine: A Theology of Health Care* (Eugene, OR: Wipf and Stock, 2008), pp. 100-116. The relationship between the healing narratives in the Gospels and Acts and texts like James 5 is, however, disputed, even among those who wish to trace fairly direct connections between the practice of healing commended in the latter and church life today. The Pentecostal biblical scholar Keith Warrington, for example, argues that the primary purpose of the healing stories of the Gospels is to demonstrate Jesus' unique authority and the inbreaking of God's kingdom in his life and ministry; the healings in Acts, likewise, are narrated first and foremost to demonstrate Christ's continuing presence in the early Church. While acknowledging that there are of course lessons for the healing ministry of today that can be learned from the ministry of Jesus, he cautions against attempting to derive a "methodology" for the healing ministry from the Gospels and Acts. James 5, in his view, offers more detailed guidelines for the Church's ministry of healing — albeit in a specific context very different from today's churches, so that James's instructions about the praxis of healing must be recontextualized if they are to guide today's healing ministry. See, e.g., Keith Warrington, "James 5:14-18: Healing Then and Now," *International Review of Mission* 93.370/371 (2004): 346-367; "Acts and the Healing Narratives: Why?" *Journal of Pentecostal Theology* 14.2 (2006): 189-217; "Healing and Suffering in the Bible," *International Review of Mission* 95.376/377 (2006): 154-164; "A Response to James Shelton Concerning Jesus and Healing: Yesterday and Today," *Journal of Pentecostal Theology* 15.2 (2007): 185-193. His fellow Pentecostal James Shelton, on the other hand, finds more continuity between the healing depicted in the Gospels and James 5, and is less wary than Warrington of using the ministry of Jesus and the Apostles as a model for contemporary practice: James B. Shelton, "'Not Like It Used to Be'? Jesus, Mira-

ical and contemporary expressions of that tradition in global Christianity are almost endlessly varied; even if we were to restrict our attention to the contemporary West, we would find enormous diversity of practice. Some structure could be lent to this diverse picture by a typology developed in the Church of England report *A Time to Heal,* which identifies "pastoral," "sacramental," and "charismatic" aspects to Jesus' healing ministry, and discerns the same three "streams" in the Church's historical healing ministry.[7] The healing stories depict Jesus' compassion for the sick, a *pastoral* concern that (the authors remark) extends beyond the people of Israel to Gentiles such as the centurion's servant (Matt. 8:13) and the daughter of the Syrophoenician woman (Mark 7:24-30).[8] This pastoral aspect has been reflected historically in the care of the sick offered by church leaders and Christians who risked their lives in times of plague, in the practice of monastic communities in caring for the sick, in the health care exercised by parish clergy in medieval and early modern times, and in longstanding links between the Church and the medical and nursing professions. The *sacramental* character of Jesus' ministry is intimately linked to the Incarnation, in which his physical humanity is central to God's saving work through the cross and the resurrection. In Jesus' ministry, physical things and acts such as the touch of his hands, the use of spittle and dust, and the command to perform a physical action such as washing in the pool of Siloam, become "effective signs" and instruments of God's healing work (e.g., Mark 1:31, 8:23; Luke 4:40; John 9:6-7). The sacramental character of Christian healing ministry has been expressed historically in its close link with the Eucharist and in the varied uses of anointing with oil and the laying on of hands. The *charismatic* aspect of Jesus' ministry is announced (for example) in his appropriation of a prophetic text from the book of Isaiah, "The Spirit of the Lord is upon me, because he has anointed me . . ." (Luke 4:18-19, quoting Isa. 61:1-2). The Acts of the Apostles and Paul's

---

cles, and Today," *Journal of Pentecostal Theology* 14.2 (2006): 219-227; "A Reply to Keith Warrington's Response to 'Jesus and Healing: Yesterday and Today,'" *Journal of Pentecostal Theology* 16 (2008): 113-117.

7. Perry, ed., *A Time to Heal,* pp. 26-34.

8. It must be said, though, that Jesus' pastoral concern for the Syrophoenician woman and her daughter is represented ambiguously (to say the least) in the perplexing and troubling dialogue that precedes the healing: see, further, Peter Admirand, "Traversing Towards the Other (Mark 7:24-30): The Syrophoenician Woman Amidst Voicelessness and Loss," in *The Bible: Culture, Community, and Society,* ed. Angus Paddison and Neil Messer (London: T&T Clark, 2013), pp. 157-170.

teaching about spiritual gifts (1 Cor. 12–14) emphasize that the healings done through the first Christians were a gift of the Spirit, and historically healings have been associated at various times with charismatically gifted individuals. As the report acknowledges, the charismatic aspect of the Church's healing ministry has at times been neglected in the West, and both Pentecostal and Orthodox spiritualities have been a salutary reminder to Western Christians that the ministry of healing reflects the Holy Spirit at work in the Church. This typology of pastoral, sacramental, and charismatic aspects of healing ministry could offer a helpful analytical lens through which to examine contemporary Christian healing practice, though it is not put to any such use in the report's own descriptive survey of the contemporary healing ministry in the Church of England and its ecumenical partners.[9]

It is beyond the scope of this chapter to survey the full range of contemporary Christian healing practice, even in one particular context. Some of this work has been helpfully done by others in various contexts. For example, *A Time to Heal* gives a detailed picture of the variety of (English) expressions of healing ministry in parishes and dioceses and at the national level, through the work of hospital, hospice, and prison chaplains, in guilds like the Guild of St. Raphael and the Order of St. Luke, and at Christian healing centers and shrines like Walsingham in Norfolk; it also surveys the practice in many of the Church of England's ecumenical partners, including Orthodox, Catholic, Protestant, and New Churches.[10] What will be done in the remainder of this section is not to repeat or extend such work, but to identify some important themes and issues arising from the literature on the ministry of healing, which have a bearing on the present project of understanding health, disease, and illness theologically.

Healing, Evangelism, and Christian Witness

It is not surprising that a good deal of the contemporary discussion of healing ministry is found in the missiology literature.[11] The link between

9. Perry, ed., *A Time to Heal*, pp. 36-88.

10. Perry, ed., *A Time to Heal*, pp. 36-88. Much of this survey is concerned with structures, systems, and organizations rather than attending in detail to liturgical or spiritual practice, though a later chapter of the report (pp. 236-57) does focus on the liturgical structure and practice of healing services in the Church of England.

11. For one example, see Jacques Mathey, ed., Theme Issue on "The Global Health Situa-

the Church's evangelistic mission and the practice of healing is clear, and is traced by practitioners back to New Testament texts such as the sending out of the disciples by Jesus (Matt. 10:1-15; Luke 9:1-6) and the close links between healing and the proclamation of the message in Acts (e.g., 3:1–4:4). Commentators such as Keith Warrington argue that one of the important outcomes of the healings in the Gospel narratives is for others to recognize Jesus' authority, and that the main purpose of the healing narratives in Acts is to demonstrate Jesus' continuing presence with the disciples.[12]

A similar connection is sometimes made in contemporary church life by regarding healing as part of the Church's evangelistic activity, so that those who witness miraculous healings will be convinced of the power of God and the truth of the gospel. This has often been a particularly strong emphasis within Pentecostalism (though even among Pentecostal authors there is some divergence about the proper balance between healing as an evangelistic activity and healing as an aspect of pastoral care *within* the community of believers).[13] Others of course are wary of this way of making the connections — particularly when it finds expression in large-scale healing rallies with an emphasis on the spectacular, which are sometimes suspected of being "manipulative and unhealthily emotional," and of leaving vulnerable people unsupported after the event.[14] But some of those who express such concerns nonetheless affirm that the ministry of healing

---

tion and the Mission of the Church in the 21st Century," *International Review of Mission* 95.376/377 (2006). See also Christoffer H. Grundmann, "Healing: A Challenge to Church and Theology," *International Review of Mission* 90.356/357 (2001): 26-40.

12. See above, n. 4.

13. See, e.g., Keith Warrington, "The Path to Wholeness: Beliefs and Practices Relating to Healing in Pentecostalism," *Evangel* 21.2 (2003): 45-49 (p. 47); Daniel Chiquete, "Healing, Salvation, and Mission: The Ministry of Healing in Latin American Pentecostalism," *International Review of Mission* 93.370/371 (2004): 474-485 (pp. 479-482). Chiquete also makes a most important connection between the importance of healing in Latin American Pentecostalism and the socio-economic context of poverty and social exclusion in many Latin American countries: "For large sections of the population in Latin America (and in the world), falling sick can become a desperate problem affecting the very basis of their existence and all areas of life, including religious faith. These people generally do not have access to state and private health care systems so that, apart from folk medicine, they have to depend on divine providence to restore them to health. Pentecostals, most of whom come from the most deprived groups in society, are also directly affected by this state of affairs. But their belief that God is a 'healing God' gives them a special kind of strength and a different perception of sickness and health" (p. 475).

14. Perry, ed., *A Time to Heal*, pp. 244-245.

is inescapably part of the proclamation of the gospel to which the Church is called and committed, because it continues Christ's mission and ministry in the power of the Spirit.[15] It enacts the proclamation of the good news of Jesus' victory in the cross and the resurrection over the forces of sin, chaos, and destruction; it witnesses to God's love and compassion disclosed in Jesus; and it gives an eschatological witness, pointing forward to God's good future in which "Death will be no more; mourning and crying and pain will be no more" (Rev. 21:4) and the leaves of the tree of life in the heavenly city are "for the healing of the nations" (Rev. 22:2).[16]

Healing and Wholeness

What do those engaged in the Christian ministry of healing think they are seeking? In much of the Christian literature, healing is understood as a quest for wholeness — often with an appeal to the Hebrew word *šālôm*, interpreted holistically as well-being of body, mind, personal and social relationships, and right relationship with God. The authors of *A Time to Heal* follow a line somewhat like this: in the course of a discussion of healing miracles, they remark that "in their New Testament context the miraculous healings were not isolated from the wider healing of relationships with God and with others."[17] Some authors, such as John Wilkinson, connect the notion of *šālôm* that they find in their biblical sources explicitly with the WHO definition ("a state of complete physical, mental, and social well-being . . ."),[18] while for others there is a strong implicit connection. For example, *A Time to Heal* quotes with approval the definition of the World Council of Churches' Christian Medical Commission: "Health is a dynamic state of well-being of the individual and society, of physical, mental, spiritual, economic, political, and social well-being — of being in harmony with each other, with the material environment and with God."[19] Such an understanding is often taken for granted in the literature on

---

15. E.g., Perry, ed., *A Time to Heal*, pp. 258-281.
16. So, e.g., Cornelius A. Buller, "Healing Hope: Physical Healing and Resurrection Hope in a Postmodern Context," *Journal of Pentecostal Theology* 10.2 (2002): 74-92.
17. Perry, ed., *A Time to Heal*, p. 210.
18. John Wilkinson, *The Bible and Healing: A Medical and Theological Commentary* (Grand Rapids: Eerdmans, 1998), pp. 11-13.
19. World Council of Churches Christian Medical Commission, *Healing and Wholeness* (Geneva: World Council of Churches, 1990), p. 6, quoted in Perry, ed., *A Time to Heal*, p. xv.

health and healing, and if correct, it would seem to require some revision of my argument in chapter 1. But how well-founded is it?

This emphasis on health as wholeness can claim some warrant in the sources and the tradition of understanding and practice that has sprung from them. It is a commonplace that in the New Testament vocabulary of healing, physical cure, the restoration of relationships, and salvation are not sharply differentiated from one another.[20] *Sōzō* and its cognates in particular seem to refer in different contexts to healing, rescue, and salvation, or sometimes perhaps more than one of these at the same time. The phrase with which the story of the ten *leproi* (Luke 17:11-19) concludes, *hē pistis sou sesōken se*, appears four times in Luke's Gospel and is variously translated "your faith has saved you" and "your faith has made you well" in the NRSV.[21] Commentators such as Gaiser and Evans argue that not only the vocabulary, but the narratives themselves, associate these various aspects of well-being quite closely. In Luke 17:11-19, for example, all ten *leproi* are "made clean" *(ekatharisthēsan)*, but only one, seeing that he has been "healed" *(iathē)*, returns to "give praise to God," and he is the one who is told "your faith has made you well" (or "saved you": *sesōken se*). In Gaiser's reading of this story, all ten receive a cure from God — which itself is not to be belittled — but only the one who returns receives full healing, which the story depicts as "a remarkably particular sense of wholeness, one that recognizes God, who created the world and who elected Israel, as a player in human life, one that sees the eschatological saving and healing presence

20. This is not to say, however, that they are completely synonymous. For example, *iaomai* was a medical term in classical Greek, and in the New Testament it is mostly used to refer to physical cures. The cognate noun is *iatros*, "physician," used of Luke in Col. 4:14. *Therapeuō*, on the other hand, was generally used outside the New Testament to denote service or attendance, including service to a god. In the New Testament it is used in this sense in Acts 17:25, but otherwise refers to healings or exorcisms — relatively infrequent uses outside the New Testament. *Sōzō*, derived from the adjective *sōs* ("safe"), denotes deliverance from a wide range of dangers, including natural hazards, disease, death, and divine judgment. See Evans, *Redeeming Marketplace Medicine*, pp. 88-90, and Wilkinson, *The Bible and Healing*, pp. 77-86.

21. The phrase is found in Luke 7:50, 8:48, 17:19, and 18:42; see further Joel B. Green, *The Gospel of Luke*, New International Commentary on the New Testament (Grand Rapids: Eerdmans, 1997), p. 627. Although the NRSV follows convention in translating *leproi* as "lepers," it is well known (and acknowledged in a footnote to v. 12) that they could have had a variety of skin conditions which would have rendered them ritually unclean, but almost certainly not what is now known as leprosy (Hansen's disease). The stigma traditionally associated with the noun "leper" means that it should be used with some caution, and I avoid it where possible.

of God in the person of Jesus."[22] The association of healing with the restoration of relationships is seen in stories such as the woman with the flow of blood (Mark 5:25-34), whose condition would have rendered her ritually unclean and whose healing therefore restored her to a place in the covenant community. Moreover, in the weaving together of that story with the raising of the synagogue ruler's daughter, some readers discern a transformation of social relationships, in that the one who takes priority is the unclean, marginalized woman rather than the child of the community leader, despite the obvious urgency of the latter's case.[23] Other narratives associate healing explicitly with the forgiveness of sins, notably the story of the paralyzed man lowered through the roof by his friends (Mark 2:1-12//Luke 5:17-26), who is told "your sins are forgiven" before the physical healing even takes place. The association between physical healing and the forgiveness of sins is also made explicit in the text from James 5 quoted at the start of this section.[24] Gaiser sums up this way of reading the biblical evidence:

> Even a quick exercise with a concordance makes clear that "healing" is understood very broadly in the Bible. To the sufferer, no doubt, the alleviation of symptoms will be the "healing" reality first sought, but this does not exhaust the Bible's definition of the term, not even for individuals.... Healing too will involve body and mind, soul, and spirit, as well as relationships and community, ethics, and justice. To be healed, finally, is to be complete *(šālôm)*.[25]

The close connections between healing, well-being, and salvation, and between Jesus as healer and savior, are continued in the Church's theological, liturgical, and sacramental traditions — as, for example, in the patristic metaphor of "Christ the physician" *(Christus medicus)*[26] and Ignatius of Antioch's description of the Eucharist as "the medicine of immortality."[27] Much later, some of the same connections are suggested by the General

---

22. Gaiser, *Healing in the Bible*, p. 188.
23. Cf. Jacques Mathey, "Opening Biblical Reflection," *International Review of Mission* 95.376/377 (2006): 151-153.
24. However, as will be noted later, other texts such as John 9:2-5 question any simple causal link between a person's sins and his or her disease; see below, pp. 129-130, and Evans, *Redeeming Marketplace Medicine*, pp. 82-83.
25. Gaiser, *Healing in the Bible*, p. 243.
26. Grundmann, "Healing," pp. 33-34.
27. Ignatius of Antioch, *Letter to the Ephesians*, 20:2.

Confession first used in Thomas Cranmer's 1552 Prayer Book, and formative of Anglican piety and prayer down to the present day:

> Almighty and most merciful Father, we have erred, and strayed from thy ways like lost sheep, we have followed too much the devices and desires of our own hearts, we have offended against thy holy laws, we have left undone those things which we ought to have done, and we have done those things which we ought not to have done, *and there is no health in us.* . . . .[28]

Abigail Rian Evans is one recent author who has argued at length for a theological understanding of health as wholeness.[29] She begins with a critique of what she describes as a "medical model" of health. Following Robert Veatch, she identifies four characteristics of this model: illness is non-voluntary and is not morally blameworthy; it is "equated with organic disease"; those with the relevant expertise to treat it are physicians; sickness is "any state that falls below an acceptable minimal societal standard of health."[30] While she affirms that the model has some important positive aspects, she attributes several serious drawbacks to it:[31] the medical model narrows the scope of healing to the curing of disease, promotes the dominance of physicians (thereby excluding other healers and disempowering patients), reduces health care to medical treatment, resulting in "inadequate health care services,"[32] and can give rise to the inappropriate medicalization of many areas of life. Technological medicine can create unrealistic expectations, the training of physicians in this style of medicine emphasizes the technical to the exclusion of the humanistic and the moral, and a system of high-tech scientific medicine generates ever-increasing health care costs. These criticisms lead her to advocate instead a "wholistic" concept of health.[33] She finds moves in this di-

---

28. "The Order for Morning Prayer Daily Throughout the Year," in *The Book of Common Prayer* (1662). Available online at http://www.churchofengland.org/prayer-worship/worship/book-of-common-prayer/the-order-for-morning-prayer.aspx, emphasis added.
29. Evans, *Redeeming Marketplace Medicine*.
30. Evans, *Redeeming Marketplace Medicine*, pp. 21-27 (pp. 22, 26).
31. Evans, *Redeeming Marketplace Medicine*, pp. 30-57.
32. Evans, *Redeeming Marketplace Medicine*, p. 40.
33. Evans uses the spelling "wholistic" to denote "integrated health care that draws on spiritual and scientific resources in the mainstream of medicine and religion," as opposed to "holistic" health care, which she says "generally eschews traditional medicine and follows more controversial health care practices"; *Redeeming Marketplace Medicine*, p. 171.

rection in the work of some philosophers and doctors, but holds that "[a] radical reconstruction of the understanding of healing and healers is needed"; she believes that "[t]he Judeo-Christian tradition and its Scriptures provide a positive and visionary direction for dealing with the more systemic problems in health care and can thus serve as a base point for any such reconstruction."[34] This leads her to advocate the kind of wholeness-oriented reading of biblical texts outlined above.[35]

Curiously, some authors associate an emphasis on healing as wholeness with a distinction between *curing* and *healing*. This distinction rests on another already encountered in chapter 1, between *disease* and *illness*. In this context, disease is understood as an objective condition susceptible to scientific investigation, whereas illness is taken to be a subjective experience of loss of health or well-being, constructed differently in different cultural contexts. Thus medical anthropologists studying healing practices cross-culturally might agree that a practice heals an illness experienced in a particular cultural context, without committing themselves to the claim that the practice cures any disease. The same distinction is sometimes also deployed by commentators on the New Testament healing narratives. John Dominic Crossan, for example, argues that "Jesus and his followers healed illness. They never, in my opinion, cured disease except when and if it happened indirectly through that former and much more important process."[36]

This distinction between healing and cure is sometimes taken, in contemporary reflection, to mean that the Christian ministry of healing is not primarily concerned with curing diseases, but with seeking well-being and wholeness. Such emphases can be found in church contexts as diverse as Latin American Pentecostalism and English Anglicanism. In the former, Daniel Chiquete endorses the development of "a holistic and relational concept of health . . . [focusing] largely on the emotional, spiritual, social, family, and church aspects, as opposed to the old reductionist approach that sees illness purely as a pathological disorder in the individual."[37] In the latter, the report *A Time to Heal* devotes considerable attention to a ministry concerned with the healing of emotions, memories, and relation-

34. Evans, *Redeeming Marketplace Medicine*, p. 67.
35. Evans, *Redeeming Marketplace Medicine*, pp. 67-116.
36. John Dominic Crossan, "The Life of a Mediterranean Jewish Peasant," *Christian Century* 108 (1991): 1194-1200 (p. 1197); see also Crossan, *The Historical Jesus: The Life of a Mediterranean Jewish Peasant* (New York: HarperCollins, 1992), pp. 336-337, and for a brief discussion Gaiser, *Healing in the Bible*, pp. 178-181.
37. Chiquete, "Healing, Salvation, and Mission," p. 482.

ships, for which counseling and psychotherapy are often the partner professions of choice.[38] When this kind of emphasis is placed on the healing of emotions and relationships, it is often said that healing and wholeness can be given by God in the absence of physical cure, and even death itself can be healing.

Clearly there are important and valuable insights in all this. The distinction between curing and healing can be an important corrective to an exclusive, mechanistic focus on disease processes and their treatment — a focus which, as Kay Toombs's phenomenological analysis of illness shows, can be profoundly alienating to patients.[39] Furthermore, in theological perspective, human *flourishing and well-being* must indeed be understood holistically. As we shall see in the next section in relation to Karl Barth's account of health, a theologically satisfactory account of human flourishing must speak first of all of right relationships with God, and secondarily of well-ordered interpersonal and social relationships, good functioning of body, mind, and emotions, the availability of sufficient resources to sustain bodily needs, richness of cultural and aesthetic experience, a positive relationship with the non-human creation, and so on. The boundaries between these aspects are highly porous: they do interact with one another in many complex ways.

However, there are also dangers in such an account. A strong emphasis on healing as the promotion of holistic well-being and on the distinction between healing and curing can too easily become a way to sidestep some of the difficult questions raised by the ministry of healing, such as its uneasy relationship with scientific medicine and the evidence for "miraculous" cures (which will be revisited below). When it does, unsympathetic critics are not slow to discern it. For example, in a *Guardian* newspaper column describing his experience of participating in an evangelistic Alpha course, the science journalist and committed atheist Adam Rutherford was particularly critical of the session on healing:

> Mike [the session leader] tells us he was cured of epileptic fits by the laying on of hands. Others join in with stories of varying degrees of affliction, from the relatively trivial to the devastating. All are replete with a crucial device: defying medical expertise . . . the Cochrane Col-

---

38. Perry, ed., *A Time to Heal*, pp. 107-126.
39. S. Kay Toombs, *The Meaning of Illness: A Phenomenological Account of the Different Perspectives of Physician and Patient* (Dordrecht: Kluwer Academic, 1992).

laboration analyzed the effect of intercessory prayer on health, and found no significant results. For example, they divine "no clear effect of intercessory prayer on death." I guess Mike might not be aware of these data. Instead he doles out the kind of get-out clause that only the most blindly faithful could: "even death can be healing."[40]

While Rutherford no doubt makes some questionable assumptions and tendentious assertions of his own, his scathing response does at any rate call for some serious reflection on whether anything is amiss with this commonly-voiced Christian view. It reinforces other concerns about the understanding of healing as wholeness, particularly when that understanding is linked implicitly or explicitly to the World Health Organization's account of health as a theory of everything.[41] In chapter 1, some difficulties with such an all-embracing notion of health were discussed, and there are also more explicitly theological reasons for taking these difficulties seriously. For example, as Stephen Lammers and Allen Verhey suggest, such a concept would make it difficult to give an account of tragic conflicts of goods in a world that is both finite and fallen, and would render Paul's experience of a "thorn in the flesh" (2 Cor. 12:1-10) — a key text for a Christian understanding of health and disease, as we shall see later in this chapter — unintelligible.[42] All this signals that the distinction between curing disease and healing illness, and the identification of healing with wholeness, might be in need of some critical scrutiny.

To take the distinction between curing disease and healing illness first: I argued in chapter 1 that a sharp distinction between disease and illness rests on a problematic modern dichotomy of fact and value.[43] When commentators such as Crossan and Pilch use that distinction as a lens through which to study the healing narratives of the Gospels, there is a danger that it will

---

40. Adam Rutherford, "Alpha Can't Heal My Skepticism," Comment is Free: CiF Belief, *The Guardian Online*, 4 September 2009. Available online at http://www.guardian.co.uk/commentisfree/belief/2009/sep/03/alpha-course-religion-christianity.

41. Although Evans is critical of the WHO definition, she only objects to its perfectionism, not to its broad scope; indeed, she explicitly rejects Daniel Callahan's criticisms (discussed above in chapter 1, pp. 2-6); *Redeeming Marketplace Medicine*, p. 33.

42. Stephen E. Lammers and Allen Verhey, eds., *On Moral Medicine: Theological Perspectives in Medical Ethics*, 2nd ed. (Grand Rapids: Eerdmans, 1998), p. 239.

43. See above, pp. 38-50; also Christopher Megone, "Aristotle's Function Argument and the Concept of Mental Illness," *Philosophy, Psychiatry, and Psychology* 5.3 (1998): 187-201, and Christopher Megone, "Mental Illness, Human Function, and Values," *Philosophy, Psychiatry, and Psychology* 7.1 (2000): 45-65.

distort as much as it clarifies, by projecting back onto those narratives a fact-value dichotomy that is both alien to the world of the text and conceptually questionable in itself. Gaiser observes that, unlike the present-day examples adduced by medical anthropologists of healing that does not involve cure,

> the full healing of the Samaritan leper [Luke 17:11-19] apparently includes cure also. In fact, this is most often the case in biblical healing stories. Although . . . the biblical words for "healing" are broad and inclusive terms, generally including much more than physical cure, they most often do envision physical cure as well.[44]

Although he is critical of the healing-curing distinction, Gaiser still wishes to understand healing as wholeness. Yet it might be asked whether the biblical and theological case for this understanding is as well-founded as is often thought. Arguably, some of the texts that are adduced to support such an understanding only do so if they are read through the lens *of* that decision. For example, as we have seen, Gaiser finds support for a view of healing as wholeness in the story of the ten *leproi*. Yet if one was not already committed to that view, the story could be read in a rather different way. "Cleansing," "healing," and "making well/saving" are indicated by three different Greek verbs, and the transition from the first two to the third is anything but automatic. Gaiser interprets this to mean that only one of the ten receives the "full healing" that includes right relationship with God. Could we not equally well infer, though, that healing and right relationship with God are *different and distinct* aspects of the in-breaking of God's reign? They are of course related — they are both aspects of God's good purposes for human creatures, announced and accomplished in Christ, and both will find fulfillment in the good future which is God's eschatological promise to humanity; but this side of the eschaton, there may be a case for distinguishing carefully between them.

There appears also to be some circularity in Abigail Rian Evans's biblical and theological case for healing as wholeness. For example, one of her

---

44. Gaiser, *Healing in the Bible*, p. 180. Pilch acknowledges that the authors and first hearers of the biblical narratives would not make the healing-curing distinction that he advocates, but argues that they would "agree essentially with its conclusions"; *Healing in the New Testament*, p. 72. But since the texts do appear to envision healing that includes physical cure, it is not clear that this is so; see Gaiser, *Healing in the Bible*, p. 181; also Shelton, "'Not Like It Used to Be'?" pp. 220-221.

objections to the medical model is that "narrowing the definitions of health and sickness reduces healing the whole person — and by extension, society — to curing a disease."[45] Of course that is true, but it can only be taken as an objection to the medical model if one has already decided that healing should *not* be limited to "curing a disease." An advocate of the medical model would presumably think that the understanding of healing *should* be so limited, and that other aspects of personal and societal well-being should *not* be brought under the umbrella of "health." Such a person's support for the medical model will hardly be refuted by pointing out that the model has those consequences. Again, in discussing the New Testament understanding of healing, Evans makes the following remarks:

> Examining the variety of words in the Christian Scriptures for healing is instructive in understanding the comprehensive nature of healing from the biblical perspective. They point to its broad meaning...
>
> Another term related to this broader understanding of healing is *metanoia* (change of heart, repentance), which is a prerequisite for forgiveness and connotes restoration to the fellowship of believers. Repentance and forgiveness of sins are linked, for example, in Jesus' last words to his disciples (Luke 24:47). Even more interesting is the movement in 2 Corinthians 7:9-11 from godly grief to repentance to salvation. So repentance leads to healing.[46]

It is not clear, however, why *metanoia* should be included in a list of "words related to healing," unless it has already been decided on other grounds that repentance is included in the meaning of healing; in which case the stipulation that it is a word related to healing cannot be said to *point to* "the comprehensive nature of healing from the biblical perspective." Neither of the texts cited mentions things that everyone would agree are related to healing: the context of 2 Corinthians 7:9-11 is Paul's troubled relationship with the Corinthian church.[47]

More generally, Evans's argument for understanding health as wholeness seems to be driven by an overly simple opposition of "medical" *versus* "wholistic" models. Certainly I agree with her that the "medical model" as

---

45. Evans, *Redeeming Marketplace Medicine*, p. 36.
46. Evans, *Redeeming Marketplace Medicine*, pp. 88-89.
47. For a discussion, see C. K. Barrett, *The Second Epistle to the Corinthians* (London: A&C Black, 1973), pp. 206-212.

she characterizes it is theologically untenable: in chapter 1, various problems with reductive models of this sort were identified,[48] and this chapter and the next will give a range of theological reasons for thinking of health less reductively. Furthermore, some of the wider connections Evans explores are informative and important — in particular, her helpful account of the complex relationship between sin, sickness, and suffering, to be revisited below.[49] But wholism is not the only alternative to a reductionist medical model. A range of more nuanced philosophical options has already been surveyed in chapter 1; this chapter and the next will show that a theological understanding developed in critical dialogue with those philosophical accounts can and should avoid the extremes of reductionism and wholism.

The Christian Practice of Healing and the Professional Practice of Health Care

The relationship between the Christian ministry of healing and scientific medicine can be uneasy on both sides. On the side of healing ministry, some Christians take their cue from the story of King Asa, castigated by the Chronicler for seeking help from physicians rather than the Lord (2 Chron. 16:11-13): they imagine that going to the doctor betrays a lack of faith in the healing power of God. Keith Warrington, for example, acknowledges that "at times in Pentecostal history, [medicine] has been viewed as an inferior form of healing or even inappropriate for a believer."[50] On the other side is the suspicion voiced by Adam Rutherford in his "Alpha Male" column quoted earlier, that faith in miraculous healing entails "defying medical expertise."[51] On this view, the ministry of healing might be not only nonsense, but dangerous nonsense, because it could lead sick people away from seeking the treatment they urgently need from properly qualified professionals.

---

48. Particularly with reference to Christopher Boorse's biostatistical theory, which Evans oddly appears to associate with philosophical moves *away* from reductive notions of health as the absence of disease: see *Redeeming Marketplace Medicine*, p. 62.
49. Evans, *Redeeming Marketplace Medicine*, pp. 76-83.
50. Warrington, "The Path to Wholeness," p. 46. He does, however, hold that this view has significantly declined, and most Pentecostals would now regard medical healing much more positively.
51. Rutherford, "Alpha Can't Heal My Skepticism."

Of course, almost every serious theological author on the healing ministry is quick to deny that there should be any such conflict. For example, Keith Warrington insists that "Pentecostalism tends not to contrast medical healing and divine healing and the former is not viewed suspiciously or negatively . . . most recognizing that all healings have divine origin and therefore that recourse to medication is appropriate for the Christian."[52] Likewise, the authors of *A Time to Heal* remark that "with most other Christians, we want to affirm that everything achieved by [the medical] profession which promotes true healing can be seen as a continuation of the ministry of Jesus Christ," and argue for cooperation between Christian ministers and health care professionals in both professional and parish settings.[53] Frederick Gaiser, discussing the verse "I am the Lord who heals you" (Exod. 15:26), argues that texts such as this exclude the work of other *divine* agents, but certainly not the work of human agency. When Isaiah prescribes a fig plaster for Hezekiah's boil (2 Kings 20:7), "[n]othing in that human or natural agency is meant to dispute the claim that Yahweh is Israel's healer. . . . Not only do God's claims not exclude human work, but also they normally and regularly are effective in and through human work."[54] Certainly Christians have no reason to understand divine and human agency as mutually exclusive alternatives. Gaiser is surely right to insist that in the Zimbabwean context with which his book begins, both the healing received by worshipers at an open-air service through prayer and the laying on of hands, and his brother's healing through surgery at a Harare clinic, were the work of God.[55]

Yet some aspects of the tension between Christian healing and medicine persist. One is the way in which the evidence for healing in answer to prayer is conceptualized and evaluated. For example, one chapter of *A Time to*

---

52. Warrington, "The Path to Wholeness," p. 46.

53. Perry, ed., *A Time to Heal*, p. 91. As already noted, however, the authors tend to assume that the role of Christian ministers — particularly in health care institutions — will be mostly concerned with patients' emotional, psychological, and spiritual needs, together with pastoral care of institutions' staff and a concern for institutional ethos. Certainly there is no reason to deny that these are important roles, to which ministers may well be called and for which they may be well equipped. Moreover, the importance of respecting the distinctive vocations and expertise of the various professions working in a complex healthcare institution is obvious. Nonetheless, there is a risk that the kind of division of labor commended by the report could reinforce the kind of distinction between healing and curing that I have already criticized.

54. Gaiser, *Healing in the Bible*, p. 33.

55. Gaiser, *Healing in the Bible*, pp. 1, 2.

Heal addresses the question, "Are you expecting the same sort of cures as sick people did in Jesus' day, and can such cures be shown to be miraculous after being subject to medical criteria?"[56] — a similar challenge to that posed by Adam Rutherford. The report offers a careful discussion of this question, but one that gives the impression of a struggle to hold together conflicting views, and perhaps a certain unease with the issue. The authors remark that "it is practically impossible to provide scientific evidence to prove without question the effectiveness of the Church's healing ministry,"[57] but they nonetheless wish to give credence to claims of clinical evidence for the beneficial effects of prayer.[58] Much of the chapter is concerned with the presence or absence of evidence for supernatural healings that defy natural explanation, yet the authors also warn against making "a sharp distinction between the natural and the supernatural" which would "imply a dangerous division between what medical science achieves and what prayer for healing achieves."[59] This last insight, more consistently followed through, could be a helpful corrective to the prevalent assumption that miraculous healing must be located in the gaps unfilled by scientific causal explanations of recovery from disease or disorder. While that assumption does not of course entail a "God of the gaps" (because it is perfectly possible to affirm that divine agency *also* operates through natural and human causes: God heals through medicine as well as miracle), the strategy of invoking divine action as a causal explanation for otherwise unexplained phenomena does seem uncomfortably close to a "God of the gaps" view. At any rate it can easily encourage the same false move as the latter, conceiving of causation univocally so that divine action is seen as one cause among others within the physical world. Once that assumption is made, reference to divine activity seems to be less and less needed as the growth of scientific understanding closes the explanatory gaps, and reliance on divine aid seems increasingly irrelevant as humans acquire increasing ability to intervene technologically in natural processes.[60]

---

56. Perry, ed., *A Time to Heal*, p. 208.
57. Perry, ed., *A Time to Heal*, p. 220.
58. Perry, ed., *A Time to Heal*, pp. 215-216.
59. Perry, ed., *A Time to Heal*, p. 218.
60. Terence L. Nichols, "Miracles in Science and Theology," *Zygon* 37.3 (2002): 703-715, offers a longer and more nuanced discussion of miraculous healing. He is rightly critical of David Hume's well-known definition of a miracle as "a violation of the laws of nature," arguing that this reflects a deist conception of the relation between God and nature: the latter is a closed mechanistic system, and God's intervention can only be conceptualized in the

Moreover, although there is every reason to reject the view criticized by Keith Warrington — that recourse to medicine represents a loss of faith on the part of Christians — there might nonetheless be an appropriate kind of Christian ambivalence towards the practice of medicine. The roots of this ambivalence are traceable back to the earliest sources of the tradition: on the one hand there is Jesus ben Sirach's praise of the physician and the pharmacist (Sirach 38:1-15); on the other, the story of King Asa cited earlier (2 Chron. 16:11-13).[61] While Frederick Gaiser denies that the latter story expresses criticism of physicians per se,[62] it does at any rate encourage readers to be aware of the limits of medical power. Faith in physicians is no substitute for faith in the Lord, even if — as Sirach later affirms (38:12) — the Lord made physicians too.

This longstanding theological ambivalence about medicine takes a distinctive turn in the new context created by modern science, as Gerald McKenny demonstrates with his analysis and critique of what he calls the "Baconian project."[63] By this he means an approach to medicine whose roots can be traced back to Francis Bacon in the seventeenth century, aspiring to free humanity from "finitude and fortune" by harnessing technology for the relief of suffering and the expansion of individual choice. McKenny emphatically does not wish to deny the benefits gained by modern medicine or retreat to some pre-technological golden age. He does,

---

same way as a watchmaker's interference with a mechanism that has already been set in motion. Nichols argues that Christians, like other theists, must by contrast regard the universe as an open system in which natural processes of cause and effect operate in a divine context, and "miracles are never *only* the activity of God, as if God in a miracle acts alone, nakedly, in place of natural causes. Theologically, a miracle, even an exalted miracle such as the resurrection, is always God working through or in cooperation with nature, not against it" (p. 710). There is much in his account that is a persuasive and helpful corrective to the kind of reductive view influenced by Hume. I suspect, though, that further exploration is needed of how we should speak theologically about the relationship between divine and natural causation. While Nichols does emphasize that God does not act "as one force alongside other physical forces" (p. 713), he is at times apt to frame the discussion as a question about "the 'causal joint' between divine and physical activity," or the "mechanism" by which that interaction takes place (p. 712). "Causal joint" language is common in the science and theology literature, but does seem to me to make it too easy to lapse into thinking of divine action as a cause of the same kind as the physical causes open to scientific investigation.

61. Cf. Gilbert Meilaender, *Bioethics: A Primer for Christians*, 3rd ed. (Grand Rapids: Eerdmans, 2013), pp. 8-9.

62. Gaiser, *Healing in the Bible*, pp. 30, 34.

63. Gerald P. McKenny, *To Relieve the Human Condition: Bioethics, Technology, and the Body* (Albany: State University of New York Press, 1997).

however, argue that the Baconian project and the standard form of bioethics that supports it obscure central issues about the significance of embodiment, finitude, and suffering for a good human life. He presents a range of critical perspectives to expose the shortcomings of the project. First, an understanding of nature that can generate a normative account of the human is set against the Baconian tendency to see the body as manipulable matter with no inherent moral significance.[64] Secondly, an understanding of medicine as a moral tradition with its own proper ends is opposed to the view of medicine as a morally neutral technology available for whatever ends its users choose.[65] Finally, McKenny turns his attention to the technological objectification of the body in modern medicine. He begins this part of his account with phenomenological analyses of the dehumanization involved in experiencing one's body as Other through illness and medical treatment.[66] However, as I noted in chapter 1, he draws on Foucault's genealogy of "biopower" to show the limitations of phenomenology: even a kind of medicine that is not dehumanizing can still be a form of disciplinary practice through which society produces the kinds of bodies it requires.[67] But genealogy, too, has its limits: McKenny argues that while it can effectively expose the formative activity of biopower, it cannot convincingly suggest what attitudes and practices *ought* to form us. He holds that Christian attitudes and practices concerning the body constitute a kind of self-formation that both exposes the "normalizing" ways in which modern societies form us, and offers a substantive alternative.[68]

None of this requires us to reinstate the simplistic opposition of medi-

---

64. McKenny, *To Relieve the Human Condition*, chs. 3, 4, with particular reference to James Gustafson and Hans Jonas.

65. Here McKenny (*To Relieve the Human Condition*, chs. 5, 6) draws on the contrasting work of Leon Kass and Stanley Hauerwas. The former, influenced by a kind of Aristotelian perspective, is more ready to see medicine as a moral tradition with the potential to identify and support its own internal goods and ends; the latter is well known for his claim that to sustain its deepest and costliest moral commitments, medicine needs "something very much like a church"; Stanley Hauerwas, *Suffering Presence: Theological Reflections on Medicine, the Mentally Handicapped, and the Church* (Edinburgh: T&T Clark, 1988), p. 65.

66. McKenny, *To Relieve the Human Condition*, pp. 184-199, drawing principally on Drew Leder (e.g., Drew Leder, ed., *The Body in Medical Thought and Practice* [Dordrecht: Kluwer Academic, 1992]) and Richard M. Zaner (e.g., *The Context of Self: A Phenomenological Inquiry Using Medicine as a Clue* [Athens, OH: Ohio University Press, 1981]). Cf. the account by S. Kay Toombs, *The Meaning of Illness*, discussed above in chapter 1.

67. McKenny, *To Relieve the Human Condition*, pp. 197-198.

68. McKenny, *To Relieve the Human Condition*, pp. 217-226.

cine to the Christian healing ministry that Warrington, Gaiser, the authors of *A Time to Heal,* and many others criticize. As McKenny makes clear, the choice for Christians is not the simple one of accepting or rejecting biomedical technology wholesale; rather, what is needed is "a process of self-formation in which subjects are formed by *both resisting and appropriating* technology in accordance with an alternative . . . discourse and set of practices."[69] The Church can surely celebrate the genuine gains of modern medicine, welcome it as an instrument of God's healing work, and seek opportunities for collaboration; it is just that where it finds medical attitudes and practices profoundly formed by Baconian aspirations, the friendship will have to be a critical one.

Christian Responses to Suffering and Evil in the Context of Healing Ministry

In John Green's remarkable novel *The Fault in Our Stars,* the central characters, Hazel Grace Lancaster and Augustus (Gus) Waters, are older teenagers who first meet at a cancer support group.[70] Hazel has "never been anything but terminal,"[71] whereas Gus has (apparently) been successfully treated. The pivotal episode of the novel is a short vacation in Amsterdam, where they become lovers, and where Augustus later discloses to Hazel that his cancer has recurred: like her, he is now terminally ill. He speaks — not for the first time — of his fear of "earthly oblivion," of being denied either a life or a death that makes a difference and serves a greater good:

> "If you go to the Rijksmuseum, which I really wanted to do — but who are we kidding, neither of us can walk through a museum. But

---

69. McKenny, *To Relieve the Human Condition,* p. 218, emphasis added.

70. John Green, *The Fault in Our Stars* (New York: Dutton, 2012). I am indebted to my daughter Fiona for introducing me to this book. Like many of the best writers of children's and young people's fiction, Green is adept at exploring both bright and dark sides of human experience and raising some of the deepest of questions — including those that are implicitly or explicitly theological — without either patronizing his readers or supplying facile answers. It is perhaps no coincidence that he is interested in such questions and uses his fiction so effectively to raise them, since, according to his video blog, he is a Christian whose past experience includes work as a chaplain in a children's hospital: John Green, "On Religion," online at http://www.youtube.com/watch?gl=GB&v=hXII8Wn8J3Q.

71. Green, *The Fault in Our Stars,* p. 166.

anyway, I looked at the collection online before we left. If you were to go, and hopefully someday you will, you would see a lot of paintings of dead people. You'd see Jesus on the cross, and you'd see a dude getting stabbed in the neck, and you'd see people dying at sea and in battle and a parade of martyrs. But Not. One. Single. Cancer. Kid. Nobody biting it from the plague or smallpox or yellow fever or whatever, because there is no glory in illness. There is no meaning in it. There is no honor in dying of."[72]

In any discussion of health and healing, it is not long before questions arise about the meaning of pain, suffering, and the other evils associated with sickness and disease. At its most basic, the question is often simply "Why?" In many cases, of course, causal explanations are not hard to give, either at the level of the particular (the origins and progress of this disease in this patient) or the general (the evolutionary development of pathogens, the molecular and cell biology of cancer, and so on). Much of the suffering associated with disease is, in one way or another, a product of the same evolutionary and developmental processes that have given rise to human and other life.[73] As became clear in Jerome Wakefield's dispute with Christopher Megone, surveyed in chapter 1, this raises a difficult question for any teleological view of health in which *good* is attributed to biological processes and outcomes. The very effects of pathogenic microorganisms that cause terrible suffering to their human hosts, for example, may well be highly effective reproductive strategies for the pathogenic species: from the latter's "point of view," they could be described in a sense as good.

Often, of course, the question is framed theologically — "Why did *God* allow this to happen?" — and this seems to sharpen and intensify Wakefield's critical question, turning it into something like David Hume's classic formulation of the problem of evil: "Is [God] willing to prevent evil, but not able? Then is he impotent. Is he able, but not willing? Then is he malevolent. Is he both able and willing? Whence then is evil?"[74] This ap-

---

72. Green, *The Fault in Our Stars*, p. 217.
73. It is also true, of course, that a great deal of disease, disorder, and injury is caused by human misdeeds — a point to which I shall return later in this chapter and in chapter 4.
74. David Hume, *Dialogues Concerning Natural Religion*, ed. Stanley Tweyman (London: Routledge, 1991), part 10, p. 157. After Darwin, of course, the answers to the explanatory questions make the problem of theodicy even sharper, because the very processes that give rise to the variety of life at which we marvel depend for their operation on death, waste, and destruction: the development of new characteristics in a population, and eventually of new

pears to invite an apologetic response: a theodicy that will demonstrate the coherence of belief in an all-powerful and perfectly good God in the face of evil and suffering, or a defense of divine justice against accusations of injustice. However, neither an explanation nor a theodicy can be more than a partial answer (at best) to this kind of "Why?" question. More than a quest for explanation or theological coherence, such questioning is likely to be a cry of anger and protest, and perhaps also — as for Gus Waters in *The Fault in Our Stars* — a hunger for meaning, a longing that our lives will not have been fruitless and that the suffering we and others experience will serve some recognizable purpose.

If our response to these questions about suffering and evil begins with the New Testament witness to Jesus Christ, it will take a somewhat different course from many mainstream theodicies. Where the suffering caused by disease or illness is concerned, the main focus of the Gospel narratives seems to be on what God *does* in and through Jesus Christ in response.[75] Most obviously, the healing stories are expressions of divine love made known in the compassion of Jesus. Several describe him as being "moved

---

species, depends on the over-production of offspring and differential survival and reproduction, so that those individuals in a population who are best adapted to the prevailing conditions are the ones most likely to pass on their genes to future generations. Furthermore, these selective processes — as I noted above — often give rise to effects or behaviors that are highly effective survival strategies for some creatures, but cause terrible suffering to others. The questions this raises about the love and justice of God have been recognized since the time of Darwin himself. In a letter to the American botanist (and Christian) Asa Gray, Darwin wrote, "I own that I cannot see as plainly as others do, and as I should wish to do, evidence of design and beneficence on all sides of us. There seems to me too much misery in the world. I cannot persuade myself that a beneficent and omnipotent God would have designedly created the Ichneumonidae with the express intention of their feeding within the living bodies of caterpillars, or that a cat should play with mice"; Francis Darwin, ed., *The Life and Letters of Charles Darwin, including an Autobiographical Chapter*, 3 vols. (London: John Murray, 1887), vol. 2, p. 312. Available online at http://darwin-online.org.uk/content/frameset?itemID=F1452.2&viewtype=text&pageseq=1.

75. This emphasis can be found in the work of theologians such as P. T. Forsyth and Karl Barth. For example, Barth remarks in the first part of his doctrine of reconciliation that "in [the] work of the justification of unrighteous man God also and in the first instance justifies Himself"; *CD* IV.1, p. 561. God has done this freely for our sakes, not out of any compulsion or obligation to do so. Likewise, P. T. Forsyth, responding theologically to the catastrophe of the First World War, "argues for the justice of God . . . by showing how God deals with evil historically and in practice": Colin E. Gunton, *The Actuality of Atonement: A Study of Metaphor, Rationality, and the Christian Tradition* (Edinburgh: T&T Clark, 1988), p. 106; see P. T. Forsyth, *The Justification of God: Lectures for War-time on a Christian Theodicy*, 2nd ed. (London: Latimer House, 1948). See further Messer, *Selfish Genes*, pp. 199-202.

with pity" or "compassion" (*splanchnizomai*, e.g., Matt. 20:34, Mark 1:41) — a verb also used of the Good Samaritan (Luke 10:33), who is not only presented in the parable as an example of the neighbors we should be to those around us, but often in the history of the text's interpretation has been identified with Jesus.[76] Another clear feature of the healing narratives is that Jesus enters into conflict with the diseases he heals: for example, he "rebukes" the fever of Peter's mother-in-law (*epitimaō*, Luke 4:39) as he rebukes demons (e.g., Matt. 17:18) and the wind and waves (Mark 4:39).[77] As will be discussed further in the next section, the healing narratives witness that God has declared war on the chaotic and destructive forces that threaten creation's well-being. That conflict, of course, finds its culmination in the cross and the resurrection, the decisive victory that anticipates God's promised good future in which suffering and death will finally be things of the past (Rev. 21:4). As the reference to the Good Samaritan suggests, therefore, the Church in its healing ministry is called to offer the same kind of practical compassionate response, a response which witnesses to God's love and pity and points forward to that promised future hope.

Questions of explanation and divine justice do arise, of course: most famously as central preoccupations of the book of Job, but also in the New Testament, for example in Jesus' perplexing and perhaps troubling exchange with the disciples concerning the man born blind:[78]

> His disciples asked him, "Rabbi, who sinned, this man or his parents, that he was born blind?" Jesus answered, "Neither this man nor his parents sinned; he was born blind so that God's works might be revealed in him. We must work the works of him who sent me while it is day; night is coming when no one can work. As long as I am in the world, I am the light of the world." (John 9:2-5)

---

76. See Ian A. McFarland, "Who Is My Neighbor? The Good Samaritan as a Source for Theological Anthropology," *Modern Theology* 17.1 (2001): 57-66.

77. This connection was first suggested to me by my former colleague Paul Middleton.

78. Part of what might seem troubling is the impression that the purpose of the man's blindness is to be a kind of homiletical aid; another troubling aspect is that in the exchanges that follow between the Pharisees and the man, his parents, and Jesus, blindness becomes a metaphor for alienation from God and a sinful refusal to believe (vv. 39-41). See, e.g., George R. Beasley-Murray, *John*, Word Biblical Commentary 36 (Waco: Word, 1987), pp. 160-161. As we shall see later in this chapter, the use of impairments such as blindness as metaphors for sin has been a cause of complaint against the Christian tradition for some theologians of disability such as Nancy Eiesland.

Even this text, however, tends to direct us *away* from what we often recognize as theodicy. The disciples seek a straightforward explanation for the man's condition, but Jesus' response unsettles any easy assumption that diseases or impairments are to be explained as divine punishments; furthermore — whatever else is meant by "so that God's works might be revealed in him" — it quickly shifts the focus onto "the works of God" and the sign of those works offered in the man's healing.[79]

When faced with the questions raised by disease, pain, and suffering, the *first* response of Christians should be one of practical compassion, which witnesses to the love of God revealed in the compassionate acts of Jesus Christ. To be sure, questions of explanation and theodicy may sooner or later have to be addressed, but even then, we should be rather careful how we go about it. Not long before his death, the philosopher of religion D. Z. Phillips remarked that "the problem of evil should be discussed with fear and trembling. This is because it is easy for us, as intellectuals, to add to the evil in the world by the ways in which we discuss it."[80] One of his principal targets was the kind of philosophical theodicy that defends belief in an omnipotent and perfectly good God on the grounds that there is no way for even an all-powerful God to create a world like ours without evil, but that the good of such a world justifies the evil. In one of the best-known such theodicies (whose origins he attributes to Irenaeus of Lyons), John Hick argues that a world in which human souls can be formed and grow towards perfection must be one in which we are placed at an "epistemic distance" from God, and in such a world there will be evil.[81] Such "only-way" theodicies may seem attractive to those wrestling with the meaning of suffering in the context of the healing ministry: for example, *A Time to Heal* alludes approvingly to Hick's "Irenaean" approach.[82]

---

79. As we have already seen, many New Testament texts do make *some* kind of connection between sin and disease — for example, in the text from James 5 quoted at the beginning of this section, and in those healing stories in which Jesus tells the sufferers that their sins are forgiven *before* they receive any physical cure (e.g., Mark 2:1-12). But John 9:3 unsettles any simplistic way of making that connection. The biblical sources of the Christian tradition suggest a complex relationship between evil, sin, and disease, which will be revisited in later sections of this chapter and in chapter 4.

80. D. Z. Phillips, *The Problem of Evil and the Problem of God* (London: SCM, 2004), p. 274.

81. John Hick, *Evil and the God of Love* (London: Macmillan, 1966).

82. Perry, ed., *A Time to Heal*, pp. 225-227. Although Hick is not named, the contrast between "Augustinian" and "Irenaean" theodicies, and the account of the latter, are clearly derived directly or indirectly from him.

However, they have their dangers. One of Phillips's complaints against them is that their logic entails deep distortions in their God-talk: God becomes "a member of a moral community he shares with us," a moral agent whose actions are to be judged according to a consequentialist moral calculation.[83] If Phillips is correct that this is where the logic of such theodicies leads, then they seem to entail — to say the least — an odd way of speaking of the God whom Christians worship.[84]

Christians attempting to wrestle with the meaning of suffering in the context of disease and healing might be well advised to focus less on philosophical arguments of the kind Phillips criticizes and more on what God in Christ has done to address evil and suffering, how that work of Christ shows God to be just, and how we ought to respond to what God has done. Such an approach is very close to what John Swinton calls a "practical theodicy."[85] Somewhat like Phillips, Swinton is highly critical of philosophical theodicies, holding that they can actually become sources of evil by seeking to explain it, justifying its existence, and silencing the voice of innocent suffering. Instead, he argues, the Church's proper response is *resistance* to suffering by means of distinctive — and sometimes, to others, scandalous — Christian practices such as lament, forgiveness, thoughtfulness, and hospitality.

The story of the man born blind in John 9 begins with a question of explanation but ends with the man's healing. Even in the New Testament, however, not everyone gets healed. For example, St. Paul's "thorn in the flesh," to be discussed further in the next section, was not taken from him despite his repeated pleas (2 Cor. 12:1-10). Granted the eschatological promise that "mourning and crying and pain will be no more," in this present life there is pain and disease that will not be cured and must simply be endured. In this respect at least, the New Testament narratives echo the experience of the modern health care professions, that however remarkable the advances of scientific medicine, there will still be patients who cannot be cured. This observation led Stanley Hauerwas in an early essay to describe medicine as a

83. Phillips, *The Problem of Evil*, p. 46.
84. For a fuller critique of "only-way" theodicies in the different — though related — context of evolutionary biology, see Neil Messer, "Natural Evil after Darwin," in *Theology After Darwin*, ed. Michael Northcott and R. J. Berry (Carlisle: Paternoster, 2009), pp. 139-154; and for a response, Christopher Southgate, "Rereading Genesis, John, and Job: A Christian Response to Darwinism," *Zygon* 46.2 (2011): 370-395 (pp. 377-384).
85. John Swinton, *Raging with Compassion: Pastoral Responses to the Problem of Evil* (Grand Rapids: Eerdmans, 2007).

"tragic profession," in the sense that its practitioners' very vocation to care for the sick inevitably faces them with limits of their ability to do so.[86]

Faced with diseases that will not be healed in this life, and suffering and dying that appear meaningless and futile, a Christian tradition might respond in distinctive and surprising ways. Consider, for example, the loss of dignity, autonomy, and control that some terminally ill patients seem to fear even more than pain — so that even when there is a good prospect that their pain can be managed by skilled palliative care, the felt need for euthanasia or assisted suicide persists.[87] It is not surprising that this should be so, when we are profoundly shaped by a culture that accustoms us to measuring our worth by our capacity to direct our own lives and make our mark on the world. In such a context, loss of autonomy and dignity and a futile death can seem like ultimate disasters. In *The Fault in Our Stars,* part of what might be described as Augustus Waters's "passion story" is the progressive loss of his dignity and any possibility of his either living or dying heroically. Yet there are hints that he comes to understand heroism differently, that his and Hazel's love for each other begins to set him free from the fear that an inglorious death will render his life meaningless. In a eulogy that he writes for her during his final weeks, which she does not receive until after his death, he observes that most people have a compulsion to leave their mark on the world "in a ridiculous attempt to survive our deaths," and very often the marks that we leave turn out to be scars. But Hazel, he says, is different:

> She walks lightly upon the earth. Hazel knows the truth: We're as likely to hurt the universe as we are to help it, and we're not likely to do either.
>
> People will say it's sad that she leaves a lesser scar, that fewer remember her, that she was loved deeply but not widely. But it's not sad. . . . It's triumphant. It's heroic. Isn't that the real heroism? Like the doctors say: First, do no harm.[88]

---

86. Stanley Hauerwas, *Truthfulness and Tragedy: Further Investigations into Christian Ethics* (Notre Dame: University of Notre Dame Press, 1977), pp. 184-202.

87. For example, the annual reports on the operation of Oregon's Death with Dignity Act list the most commonly mentioned end-of-life concerns as "decreasing ability to participate in activities that made life enjoyable," "loss of autonomy," and "loss of dignity"; Oregon Public Health Division, *Oregon's Death with Dignity Act, 2011*, p. 2. Available online at http://public.health.oregon.gov/ProviderPartnerResources/EvaluationResearch/DeathwithDignityAct/Documents/year14.pdf.

88. Green, *The Fault in Our Stars,* pp. 311-312.

A theological reader of this story might reflect that all genuine human love has its source and goal in God's love, and it is the love of God that ultimately sets us free from the need to prove the meaning and value of our lives by our own efforts. If the "parade of martyrs" in the Rijksmuseum are examples to Gus of the kind of heroic life and death of which his cancer has robbed him, then either he or their painters have misread their stories. Part of the inversion of human values brought about by the story of Jesus' life, death, and resurrection is that we are no longer called to be heroes, but saints.[89] That is why — as Jesus tells Peter at the end of John's Gospel — a distinctly un-heroic death, in which Peter stretches out his hands and is led where he does not wish to go, can be "the kind of death by which he would glorify God" (John 21:18-19). Likewise, in Dietrich Bonhoeffer's poem "Stations on the Road to Freedom," after "Action" comes "Suffering," in which

> A change has come indeed. Your hands, so strong and active,
> are bound; in helplessness now you see your action
> is ended; you sigh with relief, your cause committing
> to stronger hands; so now you may rest contented.
> Only for one blissful moment could you draw near to touch freedom;
> Then, that it might be perfected in glory, you gave it to God.[90]

It is the message of hope in the gospel that sets us free from the need to make our lives and our deaths mean something by dint of our own striv-

---

89. Samuel Wells, "The Disarming Virtue of Stanley Hauerwas," *Scottish Journal of Theology* 52.1 (1999): 82-88.

90. Dietrich Bonhoeffer, *Letters and Papers from Prison*, enlarged ed., ed. Eberhard Bethge, trans. Reginald Fuller, Frank Clark, John Bowden, et al. (London: SCM, 1971), p. 371. The poem was written in Tegel prison a few months before Bonhoeffer's execution in 1945. The German reads:

> Wunderbare Verwandlung. Die starken, tätigen Hände
> sind dir gebunden. Ohnmächtig, einsam siehst du das Ende
> deiner Tat. Doch atmest du auf und legst das Rechte
> still und getrost in stärkere Hand und gibst dich zufrieden.
> Nur einen Augenblick berührtest du selig die Freiheit,
> dann übergabst du sie Gott, damit er sie herrlich vollende.

Dietrich Bonhoeffer, "Stationen der Freiheit" (1944). Available online at http://www.luther2017.de/465-dietrich-bonhoeffer-stationen-der-freiheit. For the connection between "Stations on the Road to Freedom" and John 21:18-19, I am indebted to a sermon by Keith Clements broadcast on BBC Radio 4, Sunday, 25 March 2012; transcript online at http://www.bbc.co.uk/programmes/b01dtfsp.

ing, and it is this which can save even indignity and apparently futile suffering and death from being ultimate catastrophes. If that is so, then the Church is called to bear witness to this in its care of the sick, by refusing to abandon those who experience any extremity of suffering or indignity. Some such line of thought is what led Hauerwas to argue in the 1980s that to sustain it in its tragic vocation of caring for those it cannot cure, the medical profession needs a community very like the Church.[91]

The Account of Health in Karl Barth's Ethics of Creation

The last section began with James's promise that "[t]he prayer of faith will save the sick ... [t]he prayer of the righteous is powerful and effective," and as we have seen, the New Testament contains many stories of people who were powerfully healed by Jesus or through the ministry of his disciples. One of the most important New Testament texts for a theological reflection on health and healing, however, tells the story of a prayer of faith that was on the face of it *not* powerful or effective — the story that has given the English language the proverbial phrase "a thorn in the flesh":

> [T]o keep me from being too elated [on account of visions and revelations of the Lord], a thorn was given me in the flesh, a messenger of Satan to torment me, to keep me from being too elated. Three times I appealed to the Lord about this, that it would leave me, but he said to me, "My grace is sufficient for you, for power is made perfect in weakness." So, I will boast all the more gladly of my weaknesses, so that the power of Christ may dwell in me. Therefore I am content with weaknesses, insults, hardships, persecutions, and calamities for the sake of Christ; for whenever I am weak, then I am strong. (2 Cor. 12:7-10)

This story is shot through with ambiguity and paradox.[92] The thorn was, to say the least, highly unwelcome: it is described as a "messenger of Satan" sent to "torment" Paul (the verb is *kolaphizō*, which connotes beating or

---

91. Hauerwas, *Suffering Presence*, pp. 63-83.
92. This paragraph and the next are taken, with modifications, from Neil Messer, "Health, the Human Genome Project, and the 'Tyranny of Normality,'" in *Brave New World? Theology, Ethics, and the Human Genome Project*, ed. Celia Deane-Drummond (London: T&T Clark, 2003), pp. 91-115 (p. 94).

battering). He pleaded with the Lord "three times" to be relieved of it, but his plea was refused, and the Lord told him, "[my] power is made perfect (*teleitai:* 'completed' or 'fulfilled') in weakness." As a result of this divine answer, Paul has come to see his "weaknesses" as the things in which he should boast, "so that the power of Christ may dwell in me." He is content to experience weakness and suffering, because, as he puts it, "whenever I am weak, then I am strong." He appears to be saying at least two related things here: first, that it is his own weakness which forces him to rely most fully on the strength of Christ, and so to experience that strength most fully, and secondly, that in C. K. Barrett's words, "a display of human weakness is the best possible stage for the display of divine power."[93] But there are also echoes of the power-in-weakness of the crucified Christ himself, and of Paul's words in 1 Corinthians that "God's foolishness is wiser than human wisdom, and God's weakness is stronger than human strength" (1:25).

The ambiguity and paradox of this passage come in part from its context.[94] Paul's account of boasting in his weaknesses is the climax of his rejoinder to his opponents in Corinth, who have presumably been given to boasting of their superior spiritual credentials and achievements. The Lord's answer, that "power is made perfect in weakness," reminds us of the necessarily paradoxical nature of any Christian reflection on power and success, since such reflection must take its bearings from the cross of Christ.

While commentators are divided on whether the "thorn" was a physical illness or some other form of suffering, it seems highly plausible to think that Paul is telling the story of some physical malady.[95] If so, it is odd that Karl Barth, in one of the most influential theological treatments of health and disease in the twentieth century, does not quote or cite this text, since — as we shall see — Paul's dialectics of power and weakness, suffering and strength, are very close to the account that Barth develops. Indeed, Stephen Lammers and Allen Verhey go so far as to remark that Barth's discussion of health and disease "may be regarded as an extended commentary on these words of Paul."[96]

Barth's account is located in the context of his presentation of ethics as the command of God the Creator, the concluding part of his doctrine of

---

93. Barrett, *2 Corinthians,* p. 317.
94. See Barrett, *2 Corinthians,* pp. 288-318, for a helpful discussion.
95. For an extensive survey, see Margaret E. Thrall, *A Critical and Exegetical Commentary on the Second Epistle to the Corinthians,* vol. 2 (Edinburgh: T&T Clark, 2000), pp. 809-818.
96. Lammers and Verhey, *On Moral Medicine,* p. 239.

creation.[97] He identifies four dimensions to the liberating command of the Creator, corresponding to the four aspects of his theological anthropology:[98] freedom before God, freedom in fellowship with one another, freedom for life, and freedom in limitation. Health is discussed as part of the third aspect, freedom for life. We are called not only to live responsibly before God and in relationship with one another; the command of the Creator also summons us and sets us free "quite simply . . . to exist as a living being of this particular, i.e., human, structure";[99] to participate in the life we have been given, and to respect our own and others' lives as a loan from God. Health is "strength for human life," or the power (which is itself God's gift to us) to obey this liberating command and to *be* the creatures God means us to be, so that *respect for life* — our obedient response to the Creator's command — includes "the will to be healthy."[100]

How do we know what kind of creature God means us to be, and therefore what "the will to be healthy" will entail? First and foremost, we know what it is to be truly and fully human in the light of Jesus Christ, who is both the incarnate Son and the true human being, our representative. (It is worth noting in passing that if we learn from Jesus Christ what truly fulfilled humanity looks like, this might serve to unsettle some of our culture's most cherished notions of human flourishing — a point to which I shall return later in this chapter, with reference to theologies of disability.) Secondly we know what constitutes human flourishing through the Scriptures, which witness to Christ. But the implications of this knowledge cannot be worked out in advance from the comfort of anyone's armchair or study. What it means in practice to "will to be healthy," to answer God's liberating call to live a human life, will always be a matter of the command of God in a particular place and time. Therefore there will always be a certain provisionality to our general conclusions about health, though this does not mean that what we say about health need be arbitrary or inscrutable to human reason.[101]

---

97. Barth, *CD* III.4, pp. 356-374. The following discussion of Barth's account is a modified and expanded version of Neil Messer, "Toward a Theological Understanding of Health and Disease," *Journal of the Society of Christian Ethics* 31.1 (2011): 161-178 (pp. 167-171).

98. Barth, *CD* III.2, §§44-47.

99. Barth, *CD* III.4, p. 324.

100. Barth, *CD* III.4, p. 356.

101. A point affirmed in relation to Barth's ethics in general by (among others) Nigel Biggar, *The Hastening That Waits: Karl Barth's Ethics* (Oxford: Clarendon, 1993), e.g., pp. 25-45.

While our understanding of what it entails to will to be healthy and live a fully human life must be centered on Christ as revealed in the Scriptures, I have already observed that all kinds of human knowledge, "taken captive to obey Christ," can contribute to this understanding. For example, I noted in chapter 1 that Christopher Megone's teleological account of health identifies survival and reproduction as goods, but that Megone acknowledges the difficulty of showing *why* they are goods. Barth's account suggests a theologically-shaped understanding of human flourishing that can incorporate these insights and make sense of them, but also reframe them. Thus, in the light of the resurrection, survival is no longer of ultimate importance, because each person's future is secured not by his or her avoidance of death, but by God's defeat of death in and through Christ. Yet the Christian tradition has persisted in regarding mortal life in this world as a good, a gift from God to be received with thanks and used wisely. So martyrdom has historically been honored in the Christian tradition, whereas suicide has not. Likewise, in theological perspective procreation no longer has ultimate importance, because the future of our kind is secured not by our own reproductive success but by the "new creation" inaugurated by Christ (2 Cor. 5:17). In the life to come, as Jesus tells the Sadducees, we will neither marry nor be given in marriage (Matt. 22:30). Yet in the present life, Christians have continued marrying, having children, and receiving them into the Church by baptism. Because our future is secured by God in Christ, we can — as Stanley Hauerwas observes — understand marrying and having children no longer as necessities but as vocations.[102] Having children becomes an expression of hope and trust in God; welcoming and caring for our children becomes one of the ways in which we welcome Christ in the guise of the stranger. A theologically-shaped account of human flourishing, in short, will acknowledge but qualify and reframe the claim that survival and reproduction are goods to be pursued. Having done so, it can incorporate all kinds of scientific insights about what might promote those goods. It can also incorporate the contrasting kinds of insight and understanding offered by the testimony of those who experience various kinds of disease, injury, impairment, and disability, testimony that could turn out to supply much-needed challenges, correctives, and extensions to the picture we would get from scientific sources alone.

---

102. Stanley Hauerwas, "The Radical Hope in the Annunciation: Why Both Single and Married Christians Welcome Children," in *The Hauerwas Reader*, ed. John Berkman and Michael Cartwright (Durham: Duke University Press, 2001), pp. 505-518.

Barth says surprisingly little about what health, as "strength for human life," looks like, perhaps because he recognizes the necessarily provisional character of any talk of God's command, perhaps simply because commonsense intuitions about health were easier to take for granted in his time than in ours. He does, however, make some important general points.[103] First, he describes health as "capability, vigor, and freedom,"[104] raising the possibility of a critical engagement with insights from the "capabilities" approaches discussed in earlier chapters.[105] Second, by writing of "the integration of the organs for the exercise of psycho-physical functions,"[106] he resists splitting off "body" from "soul" and giving one or the other greater priority, or fragmenting bodily life into separate organs and functions. The will to be healthy is the will to be a whole human being. In this way, indeed, Barth does reflect Christian notions of health as wholeness. Within his overall theological perspective, however, "wholeness" is given the proper scope and limits that tend to be forgotten if we allow our vision to be shaped by the grandiose ambitions expressed by the WHO definition. Health is indeed to be understood in an integrated way, as "wholeness," but it is still one human good among others, and should not be taken to encompass the totality of human well-being, or every aspect of what it is to be a human creature.

Third, he raises the question of "what is good . . . for the soul and body,"[107] and mentions hygiene, sport, and medicine in particular as ways of promoting and fostering health. Regarding medicine, he wants both to limit and to affirm the role of doctors in enabling us to be healthy. He warns against the conceit that "the doctor is the one who really heals":[108] health, or the strength to be human, is a gift that cannot be created or procured by doctors, or by anyone else. But medical skill and knowledge can play an invaluable role in removing the obstacles that disease places in its way, or limiting the effects of those obstacles, or at least palliating the suffering that they cause. Barth's qualified affirmation of the role of doctors connects with the Christian affirmation of, but also ambivalence towards, the profession of medicine, noted in the last section.

---

103. See further also James R. Cochrane, "Religion, Public Health, and a Church for the 21st Century," *International Review of Mission* 376/377 (2006): 59-72.
104. Barth, *CD* III.4, p. 356.
105. See further below, pp. 175-179.
106. Barth, *CD* III.4, p. 356.
107. Barth, *CD* III.4, p. 360.
108. Barth, *CD* III.4, p. 361.

Fourth, there is a social dimension to health. Hygiene, sport, and medicine might have important, albeit limited, roles to play in supporting the "strength for life"; but they

> arrive too late, and cannot be more than rather feeble palliatives, if such general conditions as wages, standards of living, working hours, necessary breaks, and above all housing, are so ordered, or rather disordered, that instead of counteracting they promote and perhaps even cause illness, and therefore the external impairing of the will for life and health. . . . The principle *mens sana in corpore sano* can be a highly short-sighted and brutal one if it is only understood individually and not in the wider sense of *in societate sana*.[109]

Barth suggests that this dimension of health has radical and perhaps even revolutionary socioeconomic and political implications: "If there is no other way, [the will for health] must assume the form of the will for a new and quite different order of society, guaranteeing better living conditions for all."[110]

If health is the strength to be human, what are we to make of disease, illness, or sickness?[111] Barth finds in the Scriptures two affirmations that appear contradictory, but which he holds in creative tension: disease is to be understood as an aspect of the evil and chaos threatening God's good creation, but at the same time it can remind us of the temporal and finite character of human life and thus direct us to the hope of eternal life offered by God in Christ.

The first of these insights is implied by Jesus' resistance to the power of sickness and death, evident in the Synoptic narratives of healing, exorcism, and raising of the dead, and ultimately in the victory won through his own suffering, death, and resurrection. Barth infers from narratives such as these that "[s]ickness is one of the elements in the situation of man as he has fallen victim to nothingness through his transgression, as he is thus referred wholly to the mercy of God, but as he is summoned by this reference

---

109. Barth, *CD* III.4, p. 363.
110. Barth, *CD* III.4, p. 363.
111. Barth uses the German term *Krankheit*, which his English translators usually render as "sickness" — rather unhelpfully for our present purposes, in view of the subsequent literature in the philosophy of medicine, sociology of health, and other fields that has carefully distinguished between disease, illness, and sickness. However, it is clear enough from the context that his discussion does include what I have generally called "disease."

to hope and courage and conflict."[112] By "nothingness" *(das Nichtige),* he does not simply mean "nothing." It is his way of speaking of evil: the shadow (as it were) of creation, that which God did not will in creating all that is.[113] As such, it has a strange, paradoxical existence as God's "kingdom on the left hand," not willed or created by God but "no less bound to God and dependent on Him than the creature which He created."[114] Humanity, sinful and alienated from God, has subjected itself to nothingness, and this subjection is correspondingly also to be understood as God's judgment on us. Sickness, "the forerunner and messenger of death,"[115] must be understood as an aspect of this subjection to nothingness. This might make it seem both presumptuous and futile to resist sickness or to will to be healthy: would we not then be resisting God's judgment? If it were not for God's mercy, this would be quite true; but in fact, says Barth, "God Himself . . . has already marched against that realm on the left, and . . . has overcome and bound its forces and therefore those of sickness in Jesus Christ and His sacrifice."[116] This good news is also our summons to hope, courage, and resistance against the forces of chaos and sickness. We can be confident in resisting them because, and only because, God in Christ has already overcome them.

Barth insists that this first insight must not be set aside or weakened in any way, but he does wish to hold it in dialectical tension with a second aspect, suggested for example by texts such as Psalm 90:12 ("So teach us to count our days that we may gain a wise heart"). This second insight is that quite apart from our subjection to nothingness, human creaturely life is temporal and finite. Human finitude is not, as such, something we should regard as a tragedy or an evil, but comes from the goodness of the Creator. Barth suggests that sickness can be "the hard actuality which ushers in this genuinely liberating insight . . . the witness to God's creative goodness, the forerunner and messenger of . . . eternal life."[117]

Barth's dialectical understanding of disease could be expressed differently, using language borrowed from Dietrich Bonhoeffer, by saying that health is not an ultimate but a *penultimate* good.[118] By the "ultimate,"

---

112. Barth, *CD* III.4, p. 372.
113. Barth, *CD* III.3, pp. 289-368.
114. Barth, *CD* III.4, p. 366.
115. Barth, *CD* III.4, p. 366.
116. Barth, *CD* III.4, p. 368.
117. Barth, *CD* III.4, p. 373.
118. Dietrich Bonhoeffer, *Ethics*, Dietrich Bonhoeffer Works, vol. 6, ed. Ilse Tödt, Heinz

Bonhoeffer means God's last word to us of judgment and salvation in Jesus Christ. The ultimate pronounces God's judgment on all that has come before, on all our this-worldly efforts to reach God or earn God's favor. At the same time, however, it gives the present world a new status and validity as *penultimate:* as the place where we can encounter and receive God's salvation in Christ. To describe health as a penultimate good, then, is to affirm it as a genuine but limited good, important not for its own sake but for the sake of the ultimate. This could encourage a certain humility in the practice of health care — not, it should be emphasized, the kind of fatalistic capitulation to sickness which Barth rather sharply describes as "the mistaken and perverted humility of the devil and demons,"[119] but a proper sense of the limits of health and the will to be healthy. In a culture that is apt to place on the biomedical sciences and health care professions the unbearable burden of trying to postpone or defeat death itself and make humans into immortal, limitless beings,[120] this proper kind of humility is a much-needed corrective.

Teleology in the Thought of Thomas Aquinas

Although Barth does not spell it out, it is clear that his account of health and disease has a teleological structure: God our Creator summons us to the life for which we have been created, and health is to be understood as the strength — also the Creator's gracious gift — to answer the summons, to flourish as creatures of the kind that we have been called and commanded to be. That there is a teleological structure to the command of God is made more explicit in Barth's general ethics in the course of his account of the "content of the divine claim":

> the grace of God — wherever it is actualized and revealed — has teleological power. It is not exhausted by the fact that God is good to us. As He is good to us, He is well-disposed towards us. And as such, He

---

Eduard Tödt, Ernst Feil, and Clifford J. Green, trans. Reinhard Krauss, Charles C. West, and Douglas W. Stott (Minneapolis: Fortress, 2005), pp. 146-170.

119. Barth, *CD* III.4, p. 368.

120. See Brent Waters, "Disability and the Quest for Perfection: A Moral and Theological Enquiry," in *Theology, Disability, and the New Genetics: Why Science Needs the Church*, ed. John Swinton and Brian Brock (London: T&T Clark, 2007), pp. 201-213; also, for a survey and variety of critical perspectives, Celia Deane-Drummond and Peter Manley Scott, eds., *Future Perfect? God, Medicine, and Human Identity* (London: T&T Clark, 2006).

wills our good. The aim of the grace actualized and revealed in God's covenant with man, is the restoration of man to the divine likeness and therefore to fellowship with God in eternal life.[121]

A theological account of health, then, should be teleological in character, not primarily in order to respond to the demands of the philosophical teleologies discussed in chapter 1, but for theological reasons — though if a theological account of health informed by the command of God the Creator has a teleological structure, this should be all to the good in enabling it to engage "irenically and polemically" (to paraphrase Barth) with those philosophical accounts.[122] Another of the resources needed to articulate a theological understanding of health and disease, therefore, will be a theological account of teleology. Though this might seem a surprising move from Barth's account of health, such a resource can best be found in what is perhaps the most influential teleological account of human being and action set within a Christian theological tradition: that of Thomas Aquinas.

A surprising move, because in an earlier paragraph of Barth's general ethics, in differentiating his theological ethics from others, he has reserved his most respectful — but all the more devastatingly severe — criticism for a Thomistic form of Catholic ethics.[123] His fundamental objection is that it takes as its starting-point the metaphysics of being rather than the divine Word addressed to us in Jesus Christ. The Thomist maxim that "grace does not destroy nature but perfects it,"[124] and the relationship between reason and revelation established by Thomas's analogical relationship between divine and creaturely being (the *analogia entis*), come in for particular criticism. The Aristotelian metaphysics on which — as we shall see — Thomas draws heavily to develop his teleology is hardly likely to prove attractive to Barth. It seems fair to conclude, then, that turning to Thomas to fill out what we should understand by the teleological power of God's grace is not a move that would have commended itself to Barth.

Recent years, however, have witnessed a number of attempts at *rapprochement* between the thought of Barth and Thomas, including the kind

---

121. Barth, *CD* II.2, p. 566.
122. Cf. Barth, *CD* II.2, p. 529.
123. Barth, *CD* II.2, pp. 529-535.
124. Thomas Aquinas, *Summa Theologica*, 2nd ed., trans. Fathers of the English Dominican Province (1920), I, q. 1, a. 8, ad 2. Available online at http://www.newadvent.org/summa/index.html. Hereafter *ST*.

of constructive synthesis in the area of ethics that goes by the name of "Barthian Thomism."[125] This has been aided, on one side, by readings of Thomas that emphasize the extent to which he is engaged in a theological reworking, informed by revelation, of what he learns from his philosophical sources. For example, Eugene Rogers reduces the distance between the maxim that "grace perfects nature" and the Barthian theme that "covenant is the internal basis of creation," by arguing that "nature" should be understood in resolutely theological terms:

> we cannot understand properly functioning nature apart from grace, and we cannot recognize nature, either as it concretely subsists in the faithful or as it concretely subsists in the unfaithful, without knowing what its *end* is, a purpose that God graciously *bestows* upon it in elevating it.[126]

From the other side, the *rapprochement* has been facilitated by those who have sought to understand Barth's ethics of creation "in terms of that conformity to the given structure of their creaturely being for which God's sanctifying command liberates humans," as Nigel Biggar puts it.[127] So in an account that will not be straightforwardly either "Barthian" or "Thomist," it seems possible to seek what Biggar calls a "constructive integration" of things learned from both, and others besides.[128] However, the integration

---

125. See, e.g., Nigel Biggar, *Behaving in Public: How to Do Christian Ethics* (Grand Rapids: Eerdmans, 2011), pp. 107-112; also Eric Gregory, "The Spirit and the Letter: Protestant Thomism and Nigel Biggar's 'Karl Barth's Ethics Revisited,'" in *Commanding Grace: Studies in Karl Barth's Ethics*, ed. Daniel L. Migliore (Grand Rapids: Eerdmans, 2010), pp. 50-59.

126. Eugene F. Rogers, *Thomas Aquinas and Karl Barth: Sacred Doctrine and the Natural Knowledge of God* (Notre Dame: University of Notre Dame Press, 1999), pp. 189-190. In a somewhat different context, Simon Oliver emphasizes that Thomas's use of his philosophical sources — whether Aristotelian or Neoplatonic — was as *tools* to accomplish his Christian theological task: "Analytic Theology: A Review Article on Oliver D. Crisp and Michael C. Rea, eds., *Analytic Theology: New Essays in the Philosophy of Theology*," International Journal of Systematic Theology 12.4 (2010): 464-475 (p. 475).

127. Biggar, *The Hastening That Waits*, p. 49. In a more recent discussion Biggar has returned to this theme, arguing that "while Barth does actually have a theological theory of the human good, it is both covert and incomplete. Rather than be covert, it should be both out in the open and right up front, and in order to be complete it needs to 'annex' (or incorporate discriminately) data from human experience and the nontheological sciences." Nigel Biggar, "Karl Barth's Ethics Revisited," in Migliore, *Commanding Grace*, pp. 26-49 (pp. 41-42). See, further, Messer, *Selfish Genes and Christian Ethics*, pp. 116-120.

128. Biggar, *Behaving in Public*, p. 108 n. 1.

must come later. In order to discern what we might need to learn from Thomas to help us answer our questions about health and disease, it is first necessary to set out his account in its own terms as clearly and carefully as possible. This section, therefore, will be concerned only with expounding Thomistic teleology and exploring in general terms what it might contribute to an understanding of health and disease. The work of constructive integration will be reserved for the next chapter.

In his understanding of what we call "teleology,"[129] Thomas draws heavily on the philosophy of Aristotle, but creatively reworks what he has learned from "the Philosopher" in a Christian theological frame of reference.[130] For Thomas, human beings are substances of a particular kind,[131] and as such have *ends* — that is to say, goods to which they are naturally inclined. This is clear, for example, in his account of the natural law in question 94 of the *Prima Secundae*:

> Since, however, good has the nature of an end, and evil, the nature of a contrary, hence it is that all those things to which man has a natural inclination are naturally apprehended by reason as being good, and consequently as objects of pursuit, and their contraries as evil, and objects of avoidance.[132]

---

129. Although interest in final causes dates back to antiquity, the *term* "teleology" and its equivalents in other languages appear to have emerged only in the eighteenth century, first being attested in the work of Christian Wolff: see Monte Ransome Johnson, *Aristotle on Teleology* (Oxford: Clarendon, 2005), p. 30.

130. The following discussion draws heavily on Simon Oliver's current work on teleology, including his forthcoming paper "Aquinas and Aristotle's Teleology," *Nova et Vetera* 11.3 (2013). I am deeply indebted to Dr. Oliver for providing me with a typescript of this paper and unpublished portions of his current work on teleology. Citations of "Aquinas and Aristotle's Teleology" in the following paragraphs refer to the typescript pagination.

131. For Thomas, either an individual human being or the human species as a whole can be described in different senses as a substance, since he accepts Aristotle's distinction between "primary" and "secondary" substances: "Substance, in the truest and primary and most definite sense of the word, is that which is neither predicable of a subject nor present in a subject; for instance, the individual man or horse. But in a secondary sense those things are called substances within which, as species, the primary substances are included; also those which, as genera, include the species. For instance, the individual man is included in the species 'man,' and the genus to which the species belongs is 'animal'; these, therefore — that is to say, the species 'man' and the genus 'animal' — are termed secondary substances." Aristotle, *Categories*, 5, 2a13-18, trans. E. M. Edgehill. Internet Classics Archive. Available online at http://classics.mit.edu/Aristotle/categories.html.

132. Thomas Aquinas, *ST* I-II, q. 94, a. 2, responsio.

The ends of a substance, or the goods towards which it is inclined, are determined by the kind of thing it is: by its *nature,* in other words its *form* and *matter.*[133] An entity's "dynamic substantial form" is what makes it an entity of one kind rather than another. In Aristotelian terms, the form is the cause of the entity's movement from potential to actuality (potency to act), or towards the realization of its characteristic ends. Dynamic substantial form, in other words, is the cause of its becoming the kind of thing it has the potential to be — as, for example, when an acorn becomes an oak tree. Thus, in terms of the "Aristotelian" fourfold scheme of causation referred to in chapter 1 (material, formal, efficient, and final),[134] formal and final causes are very closely related: as Aristotle says, the form *is* the final cause.[135]

There is an important distinction between what are sometimes called "intrinsic" and "extrinsic" teleologies. Versions of this distinction can be found in both Aristotle and Thomas, but it is carefully qualified by both. "Extrinsic" teleology is the kind in view, for example, in the making of an artifact. When clay is shaped into a pot by the efficient cause of a potter's activity, the formal and final causes — the goal of making a pot and the form of the pot that she aims to make — seem to be *external* to the material on the potter's wheel; they are in her mind. If on the other hand we think of the growth of an oak tree from an acorn or the development of a human infant into an adult as the actualization of particular forms, then we are envisaging forms *internal* to those entities. In modernity, extrinsic teleology seems relatively unproblematic, whereas the notion that entities have an internal drive towards the realization of an intrinsic form seems much harder to believe. Kant, for example, regarded internal teleology as a heuristically useful shorthand explanation for natural phenomena that would more accurately and fully be explained in terms of efficient causa-

---

133. Aquinas, *ST* I, q. 75, a. 4, responsio: "in natural things the definition does not signify the form only, but the form and the matter."

134. "Aristotelian" is placed in scare quotes, because Aristotle himself does not use these terms, though they have been widely used by Aristotle's heirs and successors. What is often called the "material" cause, Aristotle calls "that out of which a thing comes to be"; he does use the term "form," by which he means "the statement of the essence," or "what it is to be something"; what is often called the "efficient" cause, he calls "the primary source of the change or coming to rest"; and what is often called the "final" cause, he describes as cause in the sense of "end or 'that for the sake of which' a thing is done"; Aristotle, *Physics* II.3, 194b15-195a3, trans. R. P. Hardie and R. K. Gaye. Internet Classics Archive. Available online at http://classics.mit.edu/Aristotle/physics.2.ii.html.

135. "[T]he form [is] the cause in the sense of 'that for the sake of which.'" *Physics* II.8, 199a32.

tion,[136] a position similar to Jerome Wakefield's reading of Aristotelian teleology.[137] However, it seems clear that Aristotle understands what I have called intrinsic teleology to be a genuine causal schema, not simply a shorthand for efficient causation; he thinks that natural entities have what Simon Oliver calls an "intrinsic receptivity" to the kinds of change or behavior that will actualize their form. Furthermore, Oliver argues that for Aristotle this applies not only to animate but also to inanimate substances: there is a kind of tendency even in inanimate entities to realize certain forms rather than others, because matter in this world is never *completely* brute, but is always already "en-formed" to some extent (so steel makes a good saw, and glass a good windowpane, but not the other way round).[138]

Even in Aristotle, then, the distinction between the intrinsic teleology of nature and the extrinsic teleology of artifice is not clear-cut. In Thomistic thought, the distinction is complicated further by a Christian doctrine of creation. For Thomas, God is not simply an Aristotelian "first unmoved mover," but the Creator of all things out of nothing, and this means that "[a]ll natural things were produced by the Divine art, and so may be called God's works of art."[139] The forms of all creatures have been given by God in creating them *ex nihilo*.[140] So the ends of all created things are given as the Creator's gift, and there is a sense in which all created things tend to God as their ultimate end.[141] However, there is a distinctive sense in which humans, as rational creatures, have God as their "last end," and to articulate this Thomas makes use of another Aristotelian distinc-

---

136. See Immanuel Kant, *Critique of the Power of Judgment*, trans. Paul Guyer and Eric Matthews (Cambridge: Cambridge University Press, 2000), p. 269, cited by Oliver, "Aquinas and Aristotle's Teleology," p. 4.

137. Wakefield, "Aristotle as Sociobiologist."

138. Oliver, "Aquinas and Aristotle's Teleology," pp. 9-10. Oliver associates this with the Aristotelian concept of "appetition" *(orexis);* e.g., Aristotle, *Eudemian Ethics* I.8, 1218a. Perseus Digital Library. Available online at http://www.perseus.tufts.edu/hopper/text?doc=Perseus%3Atext %3A1999.01.0050%3Abook%3D1%3Asection%3D1214a.

139. Aquinas, *ST* I, q. 91, a. 3, responsio.

140. See, e.g., Aquinas, *ST* I, q. 15. This theological observation, of course, is entirely compatible with an evolutionary understanding of human biological origins, since for Thomas the doctrine of creation *ex nihilo* is not only or primarily a claim about temporal priority or sequence, but about logical priority: creation would not exist if it were not continually sustained in being by the Creator.

141. Aquinas, *ST* I-II, q. 1, a. 8, responsio; Thomas Aquinas, *Summa Contra Gentiles* III.17-25, ed. Joseph Kenny, trans. Vernon J. Bourke (New York: Hanover House, 1955-57). Available online at http://dhspriory.org/thomas/ContraGentiles.htm. Hereafter *SCG*.

tion relating to final causes. In describing the cause "for the sake of which" an action is done, Aristotle distinguishes "that of which" (the action's aim) from "that for which" (its beneficiary).[142] For example, if an artist accepts a commission to paint a portrait, she will spend a good deal of time in her studio applying paint to canvas. The finished painting is the aim of this activity ("that of which"), and the subject of the portrait, who commissioned it, is its beneficiary ("that for which"); the artist's activity in painting can equally be described as "for the sake of" the finished painting and its subject. God is the last end of all created things in the first sense, but only rational creatures can *know and love* God, and so be described as beneficiaries of their movement to their ultimate end.[143]

Inanimate and irrational creatures tend naturally towards their good, but for humans as rational creatures, it is not so straightforward. Our ultimate end is the "beatific vision": the contemplation of God which is our ultimate good or happiness. Thomas holds that this goal is necessary: "If the will be offered an object which is good universally and from every point of view, the will tends to it of necessity, if it wills anything at all; since it cannot will the opposite."[144] However, *only* the perfect good of our last end, the beatific vision, is good from every point of view and therefore a necessary end in this sense. Other goods and goals are contingent — they are not all good in all circumstances — and so decisions must be made between them.[145] This means that our journeys towards our ultimate end are a matter of deliberation, requiring all kinds of decisions about the proximate goods and goals we should pursue and the means by which we should achieve them. These decisions become sedimented, so to say, into habits of a particular kind — in other words, into *virtues,* which are "operative" habits directed towards goods.[146]

Thomas makes a distinction between "connatural" human goods — those that are in proportion to what our nature can achieve — and our ultimate good.[147] He gives some well-known examples of connatural goods in his discussion of natural law:[148] in common with all substances, we nat-

---

142. See, e.g., Aristotle, *Metaphysics*, trans. W. D. Ross, XII.7, 1072b1-4. The Classical Library. Available online at http://www.classicallibrary.org/aristotle/metaphysics/index.htm.
143. Aquinas, *ST* I-II, q. 1, a. 8, responsio; *SCG*, III.24, 25.
144. Aquinas, *ST* I-II, q. 10, a. 2, responsio.
145. See Aquinas, *ST* I-II, q. 10, a. 2, responsio and ad 3.
146. Aquinas, *ST* I-II, q. 55.
147. Aquinas, *ST* I-II, q. 55, a. 1, responsio.
148. Aquinas, *ST* I-II, q. 94, a. 2, responsio.

urally seek the goal of preserving our existence; in common with all animals, we naturally seek goals such as sex and the nurture of our offspring; proper to our nature as *rational* animals, we naturally seek goods such as living in ordered societies and knowing the truth about God. The "cardinal" virtues enable us to achieve connatural goods such as these.[149] However, our ultimate end, the vision of the divine essence, is not connatural to us, because the knowledge of God is beyond the natural powers of any creature. This means that to direct us to our ultimate end of perfect happiness, we need the *theological* virtues (faith, hope, and love), which cannot be naturally acquired, but only "infused" in us by God.[150] In other words, we can only achieve our last end, or ultimate good, by means of God's grace. Furthermore, we are all sinners, heirs of original sin, and in that condition our "natural inclination to virtue" too is damaged or diminished, so that we need God's grace even to fulfill our connatural good.[151]

What are the implications of all this for a teleological account of health as human flourishing? The first thing to say is that humans are God's creatures, brought into being *ex nihilo* as creatures of a particular kind. In Aristotelian-Thomist terms, being a creature of a particular kind means having a particular dynamic substantial form — the Creator's gift to us — which makes us creatures of this kind rather than another. Since (in Aristotelian terms) "form is final cause," being creatures of this particular kind means having particular ends. In other words, to be a creature of a particular kind just is to be a creature whose good, or flourishing, consists in some things rather than others.

It might seem, though, that Thomas's teleology plays down the value of embodied life in a way that makes his account of limited value in understanding health and disease. As we have seen, he holds that our ultimate end is the vision of God, for which the intellect does not depend on the body: "Consequently, without the body the soul can be happy."[152] This is not the whole picture, however, because he also argues that there is a sense in which the body *does* belong to the perfection of human happiness. The soul, separated from the body, can enjoy the perfect happiness of the beatific vision, but until it is reunited with a perfected ("spiritual") body at the resurrection, "it does not possess that good in every way that it would

---

149. Aquinas, *ST* I-II, q. 61.
150. Aquinas, *ST* I-II, q. 62, a. 1, responsio.
151. Aquinas, *ST* I-II, q. 85, aa. 1 and 2; see also q. 109.
152. Aquinas, *ST* I-II, q. 4, a. 5, responsio.

wish to possess it"; so when the soul is reunited with the body, its "[h]appiness increases not in intensity, but in extent."[153] Regarding the imperfect happiness that can be had in this life, Thomas readily acknowledges that the body is "in a way" needed for it, because "the happiness of this life consists in an operation of the intellect," which requires "phantasms" — mental representations of objects in the external world — that cannot be acquired without the sensory organs.[154] In other words (we might say), the soul needs knowledge of the material world, which it can only obtain by means of the sensory capacities of the body. More generally, he argues that a "well-disposed body" is needed for the happiness that can be had in this life, because this happiness requires the virtues, and we "can be hindered, by indisposition of the body, from every operation of virtue";[155] and as we have seen, there are all kinds of "proximate" or "connatural" goods associated with life in this world, such as the examples Thomas gives in his account of the natural law.

A Thomist teleology, then, can give a coherent account of human life as directed towards both proximate and ultimate ends, which are identifiable as *good* insofar as they contribute to *the good of being this kind of creature*. This account leaves us with three outstanding issues of relevance to the present project. First, although it would be an oversimplification to attribute to Thomas the kind of contempt for the body with which he is sometimes taxed, it does seem fair to say that he has bequeathed to the tradition a certain ambivalence about the goods of bodily life. This is obviously problematic for a theological account of health, which is concerned precisely with the goods of embodied human life. Second is the high place Thomas gives to humanity's rational nature. Partly under the influence of Aristotle, he holds that happiness in this life consists in the "operation of the intellect," since the rational nature is what is distinctive to humans among physical creatures; and as we have seen, he believes that it is the intellect which is required for our ultimate end, the beatific vision. To be sure, like Aristotle, Thomas has a wide-ranging conception of the activity of our rational nature — so that in his discussion of the natural law, for example, the goals proper to it include social and political as well as more narrowly "intellectual" activity.[156] Nonetheless, the critical question raised

---

153. Aquinas, *ST* I-II, q. 4, a. 5, ad 5.
154. Aquinas, *ST* I-II, q. 4, a. 5, responsio.
155. Aquinas, *ST* I-II, q. 4, a. 6, responsio.
156. Aquinas, *ST* I-II, q. 94, a. 2, responsio.

for Christopher Megone's Aristotelian teleology from a disability studies perspective in the last chapter could also be put to this Thomist account. Indeed, this critical question about the privileging of rationality will be intensified by some of the theological reflections on disability surveyed in the next section.[157] Third, how should evil be understood? Influenced in part by Augustine, Thomas offers a general account of evil as a failure or privation in the realization of an entity's proper ends.[158] Perhaps unexpectedly, there may be something in common between this understanding of evil and Barth's account of *das Nichtige*, discussed earlier — the point of contact being the "family relationship" that both accounts have to an Augustinian concept of *privatio boni*.[159] Thomas's account could offer a way of understanding diseases and illnesses as evils, and could also help distinguish between disorders and diversity: the former are states or conditions that interrupt the realization of a good that is proper to this kind of creature, whereas the latter are not. On such an understanding of diseases and other disorders, statistical normality does not have to bear anything like the weight that it does in other accounts such as Boorse's biostatistical theory of health, for which the latter has been taken to task by critics such as Ron Amundson.[160]

---

157. However, other aspects of Thomas's thought — in particular, his brief considerations of *furiosi* (those with mental disorders) and *amentes* (those with profound cognitive impairments) — might tend to deflect this critique somewhat. He holds that those who, from birth, have never had the use of reason should be baptized for the same reasons for which the Church baptizes infants; those who have some use of reason or have lucid intervals, and express a desire for baptism, should be baptized (*ST* III, q. 68, a. 12). Those who lack the use of reason cannot acquire the moral virtues, but can receive infused theological and moral virtues through baptism (e.g., *ST* I-II, q. 113, a. 3). In principle those who have never had the use of reason ought not to be given the Eucharist, but those who have limited reason and those who have had the use of reason in the past ought not to be denied this sacrament if they show, or have shown, signs of devotion towards it (*ST* III, q. 80, a. 9). See further Miguel J. Romero, "Aquinas on the *Corporis Infirmitas*: Broken Flesh and the Grammar of Grace," in *Disability in the Christian Tradition: A Reader*, ed. Brian Brock and John Swinton (Grand Rapids: Eerdmans, 2012), pp. 101-151.

158. Aquinas, *ST* I, q. 48.

159. Augustine, *The Enchiridion*, ch. 11, trans. J. F. Shaw, in *Nicene and Post-Nicene Fathers*, series 1, vol. 3, ed. Philip Schaff (Grand Rapids: Eerdmans, 1956), p. 491. Available online at http://www.ccel.org/ccel/schaff/npnf103.pdf.

160. Amundson, "Against Normal Function."

Theologies of Disability

In the first section of this chapter, I drew on the argument of Stanley Hauerwas that the practices of the Christian community give it a distinctive moral vision, which offers resources that are indispensable in supporting the medical and Christian vocation to continue caring even when we cannot cure. A related area in which Hauerwas has long emphasized the surprising distinctiveness of a moral vision formed by the Church's characteristic practices is its response to impairment and disability, particularly intellectual impairment. Since the 1970s, he has drawn attention to the ways in which intellectually impaired people call into question liberal societies' assumptions about human worth and suffering, and has argued that the distinctive practices of Christian communities should offer a radically different vision of the value of human lives.[161]

One brief example will illustrate the point.[162] Some years ago in their book *Resident Aliens*, Hauerwas and William Willimon told the story of Dorothy, who had been the Sunday school teacher's assistant in her local church for as long as anyone could remember. She was such a valued member of the church community that when she died, the whole church turned out for her funeral. The world outside the church saw Dorothy as "afflicted" with Down syndrome. The church regarded her very differently: not as "afflicted" or "unfortunate," but as a valued member of its community. It was able to do so because its moral vision was formed by narratives such as the episode in St. Matthew's Gospel in which the disciples ask Jesus "Who is the greatest in the kingdom of heaven?" and he responds by placing a child — someone marginalized and apparently insignificant — in the midst of them (Matt. 18:1-4). "In placing Dorothy, someone quite insignificant and problematic for the world, in the middle of the third grade Sunday school class," Hauerwas and Willimon remark, "Buncombe Street Church was reenacting Matthew 18:1-4 and practicing ethics in the ordinary, unspectacular yet profound and revolutionary way the church practices ethics."[163] Stories like this suggest that attention to Christian practice

---

161. See, e.g., Stanley Hauerwas, "The Gesture of a Truthful Story," *Theology Today* 42.2 (1985): 181-189; "Abled and Disabled: The Politics of Gentleness," *Christian Century* 125.24 (2008): 28-32.

162. This paragraph is taken, with modifications, from Messer, "Toward a Theological Understanding of Health and Disease," p. 172.

163. Stanley Hauerwas and William H. Willimon, *Resident Aliens* (Nashville: Abingdon, 1989), pp. 93-97 (p. 97).

might serve to destabilize dominant perceptions of normality, health, and flourishing. In this respect it might reinforce what we learn from a Christocentric perspective on health of the kind articulated earlier with reference to Barth. Furthermore, the Church's distinctive way of life *ought* to make it particularly attentive to the voices of those on the margins of human society — such as those who experience prejudice, discrimination, or exclusion as a result of impairments.[164]

"Ought," because as Nancy Eiesland argues in her pioneering and influential liberationist theology of disability,[165] too often the Church has not only failed to speak out for disabled people, but has actually contributed by its proclamation and practice to their stigmatization and exclusion. At the most practical level, church buildings and worship have often been inaccessible to people with various kinds of impairment. In addition, Christian proclamation and reflection have taken up the negative language about impairment found in some biblical texts, using conditions such as blindness and deafness as metaphors for sin, spiritual failure, and alienation from God. Furthermore, Christian practice and reflection in relation to healing has at times had a negative impact on disabled people, when impairments have been seen as divine punishments or the results of sin, or it has been assumed that if people are not healed of their impairments it is because they lack faith.[166] Thomas Reynolds argues that in theologically stigmatizing disability in such ways, the Church allows itself to be influenced by what he calls "the cult of normalcy": a complex set of disciplinary processes by which contemporary societies produce the economically productive bodies that they need, and create the unsatisfied desires that will stimulate ever-increasing consumption of goods and services. Such processes, he argues, marginalize those whose bodies are not productive in the ways that the economy requires, or who do not fit the ideals of youth, health, and beauty to which consumers are encouraged to aspire.[167] However, as well as stigmatizing disability, the Church is sometimes guilty of trivializing it by regarding disabled people as the passive objects of pity or charity, or romanticizing their lives as a source of inspiration for non-disabled people, or

---

164. See Mary Jo Iozzio, "Genetic Anomaly or Genetic Diversity: Thinking in the Key of Disability on the Human Genome," *Theological Studies* 66 (2005): 862-881.

165. Nancy L. Eiesland, *The Disabled God: Towards a Liberatory Theology of Disability* (Nashville: Abingdon, 1994).

166. Thomas E. Reynolds, *Vulnerable Communion: a Theology of Disability and Hospitality* (Grand Rapids: Brazos, 2008), pp. 34-38.

167. Reynolds, *Vulnerable Communion*, pp. 56-98.

representing them as virtuous sufferers, "those whom God favors with special burdens aimed at enhancing their spiritual capacities."[168]

In response to such complaints, Eiesland calls for the development of a liberatory theology which will combine political action with the re-imagination of Christian symbols; that re-imagination is essential, because social change requires not only practical action but also the transformation of the symbolic system that reflects and reinforces the current social order. As Reynolds points out, a central aspect of such re-imagination will be a "critical hermeneutic of disability" involving "a careful juxtaposition of texts which are themselves polyphonic and at times contradictory, so as to negotiate theologically between them and discern possible routes of fruitful analysis."[169] The reconceived symbol on which Eiesland focuses attention is "the image of Jesus Christ, the disabled God,"[170] an image that she develops into a contextual Christology. Referring to the New Testament resurrection narratives in which the disciples encounter the risen Jesus with the wounds of his crucifixion still visible, she writes,

> Here is the resurrected Christ making good on the incarnational proclamation that God would be with us, embodied as we are, incorporating the fullness of human contingency and ordinary life into God. . . . Jesus, the resurrected Savior, calls for his frightened companions to recognize in the marks of impairment their own connection with God, their own salvation. In so doing, this disabled God is also the revealer of a new humanity. The disabled God is not only the One from heaven but the revelation of true personhood, underscoring the reality that full personhood is fully compatible with the experience of disability.[171]

Eiesland suggests that this image of Jesus as the disabled God has some far-reaching theological implications. Among other things, it repudiates the notion that disability is a consequence of the disabled person's sin, and challenges our images of divine power and transcendence. It "defines the

---

168. Reynolds, *Vulnerable Communion*, pp. 38-43, quotation at p. 41. As Reynolds points out, the story of Paul's thorn in the flesh (2 Cor. 12:7-10) is sometimes used in this way; as should be clear from the discussion of that text above (pp. 134-135), using it to support such a view of disability amounts to a seriously oversimplified reading.
169. Reynolds, *Vulnerable Communion*, pp. 34-35.
170. Eiesland, *The Disabled God*, p. 90.
171. Eiesland, *The Disabled God*, p. 100.

church as a communion of justice,"[172] and calls for a rethinking of the Church's practice — centrally its liturgical practice — to make its worship, life, and faith accessible to people with disabilities. In relation to the questions about the goods, goals, and ends of human life in view in this book, it critiques a kind of Christian faith that is only capable of speaking of hope and salvation for disabled people in terms of miraculous cure, and emphasizes that for all of us, "able-bodied" or disabled, embodied existence is inevitably "subject to contingency."[173]

However, as Medi Volpe observes, several more recent authors have found Eiesland's liberationist approach with its emphasis on rights indispensable but not sufficient for Christian reflection on disability. One reason for this is that a good deal of recent work has focused in particular on intellectual impairments, which seem to bring the limits of rights discourse into sharper focus than many physical impairments. As Volpe puts it, "whereas the language of rights predicates equality on a shared human agency, which allows 'us' (the marginalized) to join 'them' (the privileged) . . . intellectual disability forces us to think differently about what makes both 'us' and 'them' human."[174] Hans Reinders, for example, acknowledges the gains that the liberal, rights-based project of inclusion has made possible for disabled people. However, the project of inclusion has its limits, because it "describes people with disabilities as self-determining individuals who with sufficient supports are perfectly capable of living their own lives according to their own preferences."[175] Those who do not have the potential to become self-determining individuals — in particular, those with profound intellectual impairments — seem to be marginalized in, or excluded from, this description. Thus, ironically, the project of inclusion turns out to be exclusionary. Other authors such as Thomas Reynolds and Amos Yong also find disability rights discourse invaluable in drawing attention to the dynamics of disabled people's exclusion, but the limitations of rights-talk compel them to draw as well, in various ways, on the more

---

172. Eiesland, *The Disabled God*, p. 104.
173. Eiesland, *The Disabled God*, p. 104.
174. Medi Ann Volpe, "Irresponsible Love: Rethinking Intellectual Disability, Humanity, and the Church," *Modern Theology* 25.3 (2009): 491-501 (p. 493).
175. Hans S. Reinders, "Understanding Humanity and Disability: Probing an Ecological Perspective," *Studies in Christian Ethics* 26.1 (2013): 37-49 (p. 42). I am most grateful to Professor Reinders for supplying me with a pre-publication copy of this paper. See further Hans S. Reinders, *Receiving the Gift of Friendship: Profound Disability, Theological Anthropology, and Ethics* (Grand Rapids: Eerdmans, 2008).

Theological Resources for Understanding Health and Disease 155

ecclesially-centered ethic of Hauerwas.[176] Such literature suggests that what is needed is a *mutually* critical dialogue between the diverse experiences of disability, the praxis of the disability rights movement, and the distinctive theological vision and practice of the Church.

Earlier sections of this chapter surveyed some resources for a theological account of health and disease that can be found in the work of Karl Barth and Thomas Aquinas. From Barth we learned that health is the "strength for human life": the God-given ability to answer the summons to flourish as an embodied creature of this particular, human, kind. As such it is *one* aspect among several of human flourishing, distinguishable (if not fully separable) from other aspects such as human relationships. The teleology implicit in Barth's account was drawn out and discussed more fully with reference to Thomas's teleological understanding of human being. So what would a mutually critical conversation about disability contribute to, or challenge in, these aspects of my account?

First, such a dialogue could lead us to think differently about theological anthropology and what it means to bear the image of God. According to Medi Volpe, an encounter with disability (particularly intellectual disability) can be a reminder that our humanity is not something we achieve by our own efforts, but a gift that we receive: "what makes us human is something that has to do with who we are before God, not who we appear to be to one another."[177] Hans Reinders is one theologian who has argued such a point in his book *Receiving the Gift of Friendship,* where he criticizes the emphasis on self-determination often found in the disability rights movement. He is also critical of official Roman Catholic teaching for its commitment to an Aristotelian-Thomist anthropology in which the end of human beings consists in the fulfillment of their rational nature — that is, their capacities of reason and will.[178] He argues that this position leads to a contradiction in Catholic anthropology, since Catholic teaching affirms the personal status of all human individuals from the moment of conception,[179] yet also seems

---

176. Reynolds, *Vulnerable Communion,* pp. 80-88, 102-132; Amos Yong, *Theology and Down Syndrome: Reimagining Disability in Late Modernity* (Waco: Baylor University Press, 2007), chs. 4, 7.

177. Volpe, "Irresponsible Love," p. 498.

178. Reinders, *Receiving the Gift of Friendship,* pp. 88-122.

179. For example, John Paul II, *Evangelium Vitae* (1995), §§60, 63. Available online at http://www.vatican.va/holy_father/john_paul_ii/encyclicals/documents/hf_jp-ii_enc_25031995_evangelium-vitae_en.html#-1M. In the context of abortion and the moral status of embryos and fetuses, §60 resists any distinction between human beings and human

committed to saying that some human individuals lack any possibility of realizing the capacities characteristic of the human kind. Reinders's proposed solution in *Receiving the Gift of Friendship* is to understand the end of human being in terms of eschatological promise, as a divine gift of friendship that does not depend upon even the potential presence of natural human capacities.[180] Some of his Catholic critics have disputed his analysis, arguing that it misrepresents the anthropology found in offical Catholic teaching.[181] Whoever is right about that, the critics have prompted Reinders to modify his own position, on the grounds that his earlier wholly eschatological solution to the exclusion of people with profound impairments left no room for an account of their embodied lives as God's *creatures*.[182] In a recent paper he argues instead for an account developed from the perspective of the doctrine of creation. In this perspective, "the purpose of human being ... is no other than *to be*."[183] Creation's purpose is to glorify God, and creatures glorify God by simply existing and flourishing.

However, the question of what constitutes specifically *human* flourishing remains to be addressed. I have argued for an account in which part of what is meant by "human creature" is a creature whose good consists in the fulfillment of certain goods rather than others: some goods, goals and ends are simply given in the structure of human creaturely life, and to flourish as a human being is to realize those goods more fully. This view may not commend itself to Reinders, who might suspect it of reintroducing the "substantialist" claim that human being depends on the possession of certain capacities — a claim he is determined to avoid. His own answer is to read the *imago Dei* in terms of "communal relations" rather than "individual substance," interpreting it to mean that God "creates a people for himself," a people that includes all God's children, and "[i]n God there is no difference in how he relates to each and every one of them."[184]

---

persons, while §63 describes the "eugenic intention" of prenatal diagnosis with a view to the abortion of fetuses with inherited disorders as "shameful and utterly reprehensible, since it presumes to measure the value of a human life only within the parameters of 'normality' and physical well-being, thus opening the way to legitimizing infanticide and euthanasia as well."

180. Reinders, *Receiving the Gift of Friendship*, pp. 279-311; cf. "Understanding Humanity and Disability," p. 43.

181. E.g., James Mumford, Review of Hans Reinders, *Receiving the Gift of Friendship*, *Studies in Christian Ethics* 23.2 (2010): 216-219.

182. Reinders, "Understanding Humanity and Disability," p. 47.

183. Reinders, "Understanding Humanity and Disability," p. 48, emphasis original.

184. Reinders, "Understanding Humanity and Disability," pp. 48-49.

I believe Reinders is quite right to read the *imago Dei* in terms of communal relations and to eschew an account of human being in which inclusion in the human community is dependent on the possession of certain capacities. However, I find it hard to see how we can avoid some kind of "substantialism" (albeit an appropriately chastened and theologically-determined kind) if we are to address the question of *what* human flourishing consists in, specifically enough to be able to discern when one of our human sisters or brothers is *not* flourishing. In other words, to be capable of recognizing threats to human flourishing and understanding what needs to be done to resist them, we need to be able to say what it means for a creature of this particular (human) kind, as opposed to some other, to flourish.

Should impairments, though, be counted among the threats to human flourishing? As we saw in chapter 2, the social model of disability is sometimes taken to mean that they should not. Yet I have argued, in common with a number of disability studies authors, that this view is overly simple. The medical model of disability as personal tragedy is inadequate, to be sure; but so is any simplistic version of the social model that obscures the experience of disability as a "complex predicament" (to use Tom Shakespeare's term) with medical, psychological, and social aspects. Some impairments can and do threaten human flourishing. However, a caveat must immediately be entered, and that leads on to a second point raised by this critical conversation about disability.

That point is that there are more ways of living a flourishing human life than we tend to imagine. For example, Hauerwas and Willimon's story of Dorothy calls into question the notion that Down syndrome should be considered a form of "weakness" opposed to the "strength for human life." Here, what Buncombe Street Church learned by being a church, listening to the Scriptures, and having Dorothy in their midst as they gathered around the Lord's table comes into conversation with Amos Yong's argument that intellectual disabilities are social constructions. Yong has no wish to deny "the very real differences between those with and without intellectual disability"; the point is that "intellectual disability is *also* a theoretical construction in terms of the prejudices against persons in that category that are embedded in our cultural systems."[185] Among other things, attention to such examples of Christian practice should seriously problematize the simple assumption that Down syndrome is properly considered a disease to

---

185. Yong, *Theology and Down Syndrome*, p. 111, emphasis in original.

which the appropriate response — in the absence of a cure — is medical intervention to prevent the birth of such children.[186]

Expressing the point more generally, such encounters with disability and attention to Christian practice draw attention to some characteristically modern distortions in our understanding of human flourishing. For example, we are sometimes too ready to assume that a flourishing human life is individual, autonomous, and independent, and that human fulfillment is somehow undermined by vulnerability and dependence. This assumption is severely criticized by authors such as Reinders and Thomas Reynolds, who trace it in part to liberal democracy's need for equal, autonomous, rational agents to participate in the social contract, and in part to the market economy's need for economically productive bodies.[187] Reynolds takes Christian churches to task for not only failing to critique the "cult of normalcy" fostered by these needs, but actually colluding with it. Christian faith too comes to be seen as an individually-chosen means to satisfy the preferences of religious consumers. "This line of thinking," he holds, "mirrors the restlessness of our commodity-driven culture. But it pulls us apart from each other and throws us back on ourselves — free, equal, independent, and rationally self-interested consumers and producers of goods."[188] The cult of normalcy not only excludes disabled people, it is destructive for all of us and for the Christian community. In opposition to the norm of independent rational autonomy, Reynolds makes vulnerability and dependence central themes in his theology of disability.[189] All human lives are characterized by these attributes: "[t]o exist as a finite creature is to be contingent and vulnerable."[190] We should not wish it any other way, since in the Incarnation, Christ himself has taken vulnerable and dependent embodied human life upon himself.[191]

---

186. Cf. Yong, *Theology and Down Syndrome*, pp. 63-66. For a powerful reflective narrative, born of the experience of having a son with Down syndrome, which problematizes such simple assumptions, see Brian Brock and Stephanie Brock, "Being Disabled in the New World of Genetic Testing: A Snapshot of Shifting Landscapes," in *Theology, Disability, and the New Genetics: Why Science Needs the Church*, ed. John Swinton and Brian Brock (London: T&T Clark, 2007), pp. 29-43.

187. Reinders, "Understanding Humanity and Disability"; Reynolds, *Vulnerable Communion*, pp. 73-101.

188. Reynolds, *Vulnerable Communion*, p. 98.

189. See especially Reynolds, *Vulnerable Communion*, pp. 102-135.

190. Reynolds, *Vulnerable Communion*, p. 18.

191. Caution is needed, however, in making the theological connection that Reynolds goes on to make: "here God reveals the divine nature as compassion not only by undergoing

In similar vein, John Swinton reflects on the participation of people with profound and complex intellectual disabilities in worship and the Eucharist. "[K]nowledge of God," he writes, "has to do primarily with participation in and response to God [who] is, in a very real sense, unknowable apart from what God chooses to reveal to human beings."[192] This means that striving for intellectual knowledge *about* God can actually be a distraction from the knowledge *of* God which comes to us as gift — as the apophatic tradition of Christian spirituality has long known. "It may be," Swinton remarks, "that our encounters with people whose life experiences include profound and complex needs remind us of truths that are often forgotten in a culture that prizes human activism and progress above most things."[193]

---

or suffering with human vulnerability, but also by raising it up into God's own being"; *Vulnerable Communion*, p. 19. He follows Moltmann in rejecting classical notions of divine impassibility, on the grounds that "like Aristotle's unmoved mover, such a God would be imprisoned in solitariness, alone in infinite perfection and unaffected by anything else. That God, then, could only appear to love, to be in relation; its being enclosed in self-absorbed contemplation, an absolute egocentricity. Neither personal nor free, that God is, in final analysis, unloving" (pp. 165-166). Against this view, he emphasizes that God's power is a "strange power," made perfect in the weakness of the cross (p. 165, quoting Daniel L. Migliore, *Faith Seeking Understanding: An Introduction to Christian Theology* [Grand Rapids: Eerdmans, 1991], p. 52). As Colin Gunton points out, however, there would be theological dangers in *over*emphasizing divine suffering in opposition to such an abstract notion of impassibility: "simply to leave the matter with a statement that God shares our suffering runs the risk of affirming suffering, making it in some way the will of God. The point of the exercise, rather, is to remove suffering from the creation, not to affirm it or establish it as in some way a necessity for God or man"; *Act and Being: Towards a Theology of the Divine Attributes* (London: SCM, 2002), p. 130. As Gunton argues (*Act and Being*, pp. 125-132), a properly Trinitarian understanding of the divine attributes can enable us to give an account of the cross as both passion *and* the exercise of divine sovereignty for our salvation, avoiding the opposite distortions of speaking of God as a solipsistic unmoved mover or as passive, at the mercy of events.

192. John Swinton, "Known by God," in *The Paradox of Disability: Responses to Jean Vanier and L'Arche Communities from Theology and the Sciences*, ed. Hans S. Reinders (Grand Rapids: Eerdmans, 2010), pp. 140-153 (p. 152).

193. Swinton, "Known by God," p. 148. However, without wishing to deny this point, Thomas Reynolds draws attention to a danger also noted by other authors: that of "trivializing disability as an instructive tool" (*Vulnerable Communion*, p. 117). There is a fine but crucial distinction between having the humility to learn from the lives of people with profound impairments, as Swinton advocates — a radical move in a culture which is accustomed to regard them as more or less purely passive recipients of care — and valuing people with such impairments *for* the lessons they can teach the rest of us.

Finally, this conversation about disability might both reinforce and amend the eschatological perspective articulated earlier in the chapter with reference to Barth. The latter, it will be recalled, holds that disease is an evil to be resisted, yet also that it can remind us of our mortality and warn us against investing our ultimate hope in this mortal life, directing us instead to the hope of eternal life offered by God in Christ. It can, in the words of the Psalmist, "[s]o teach us to count our days that we may gain a wise heart" (Ps. 90:12).[194] Theological reflection on disability can remind us to keep this eschatological horizon clearly in view in our talk of health and disease, while also raising critical questions about the ways in which the tradition conceives our eschatological hope. Amos Yong, for example, follows Eiesland and others in critiquing historical and contemporary eschatologies in which the features of disability will have no place in the resurrection life.[195] In conversation with Paul in 1 Corinthians 15 and Gregory of Nyssa's doctrine of *epektasis* (or everlasting progress), Yong proposes a "more dynamic eschatological vision" that will take account of critical perspectives from the experience of disabilities. In this dynamic eschatology, there is continuity between our present and resurrected bodies — so that at the resurrection we will recognizably be ourselves, marks of disability included — but there are also discontinuities beyond our capacity to imagine. However, in Yong's admittedly speculative proposal, "the redemptive transformation that occurs with the resurrection is only the beginning rather than the end of the soul's eschatological journey with and to God," so that "disability will be transformed even if its particular scars and marks will be redeemed, not eliminated."[196]

All this serves to remind us that for all of us — whether presently healthy or sick, presently "abled" or disabled — the ultimate fulfillment of our proper goods, goals, and ends is an eschatological promise. Our health and embodied creaturely flourishing in this life are *penultimate,* not ultimate, goods. This of course means that they are real and important goods to be fostered and protected; but also that they point beyond themselves to the ultimate fulfillment of our creaturely good in union with the triune God, our Creator, Reconciler, and Redeemer.

---

194. Barth, *CD* III.4, p. 373.
195. Yong, *Theology and Down Syndrome,* pp. 259-267.
196. Yong, *Theology and Down Syndrome,* pp. 279, 281.

Conclusion

In this chapter I have surveyed four kinds of theological resources that can contribute to a Christian understanding of health, disease, and illness: Christian practices of healing and care of the sick and theological reflection on those practices, the account of health and disease developed by Karl Barth in his ethics of creation, the teleological understanding of human life offered by the Thomist tradition, and theological reflections on disability. While I have traced some connections between these four kinds of resource, my main aim has been to present each in its own terms. In the next chapter, I shall take up the task of synthesizing insights from these theological resources, in dialogue with the philosophical and disability studies discussions surveyed in the first two chapters, to develop a unified theological account of health and disease.

CHAPTER 4

Theological Theses concerning Health, Disease, and Illness

Introduction: The Story so Far

In previous chapters, diverse areas of academic discussion and practical experience have been surveyed that have suggested insights and issues to be considered in developing an understanding of health, disease, and illness. In chapter 1 I explored the wide range of theories of health discussed in the philosophy of medicine. From this extensive discussion, various insights and issues emerged with which a theological account of health must engage. Theologians, I suggested, must be able to speak of health in ways which can critically appropriate both the teleological understanding articulated by Christopher Megone and the evolutionary insights on which Jerome Wakefield draws. We must be able to give an account of health as an aspect of human creaturely flourishing, to which both Lennart Nordenfelt's account of health as the ability to realize vital goals and the capability approach associated with Sen, Nussbaum, and others might have insights to contribute. Sensitivity to the lived realities of frailty and suffering, and to the contrasting perspectives of clinician and patient, should characterize our account. Finally, a theological account of health must be developed in a mutually critical dialogue with practices of healing and care of the sick.

Next, in chapter 2, I identified some critical insights and questions that should be put to theories of health and disease from the perspective of disability studies. Are there kinds of failure to flourish which we mistakenly attribute to individual and medical, rather than social, political, and economic causes; and if so, do we use the wrong remedies to address those failures to flourish? Do we have too narrow a view of the range of ways in which embodied human creaturely life can flourish, wrongly identifying diversity as pathology? If so, is there a danger that some interventions intended to promote flourishing will actually inhibit it for some people, as critics of the "normalization" of disabled people argue? More generally, who has the authority to say what bodily or mental conditions should be counted as harmful — for example, whose estimates of disabled people's quality of life should be taken most seriously? And is there a danger that both evolutionary accounts of natural function and teleological accounts of the human good might be ideological justifications for the operation of what Michel Foucault called "biopower," or might be distorted into such ideological justifications?

In chapter 3, I surveyed four very different kinds of resources for understanding health and disease theologically: the Christian practice of healing and care for the sick, the account of health and disease located in Karl Barth's ethics of creation, Thomas Aquinas's teleological understanding of human being, and recent theological reflections on disability. In this chapter I shall attempt to draw these diverse themes and insights together into a constructive theological account of health, disease, and illness, in critical dialogue with the philosophical and disability studies perspectives surveyed in earlier chapters. My account takes the form of sixteen theses, each with a relatively brief explanation and defense, grouped under four headings: first, humans as creatures; second, health and creaturely flourishing; third, disease, suffering, evil, and sin; finally, practical implications.

Humans as Creatures

(i) *A theological account of health and disease must begin by recognizing that we are creatures of a particular, human, kind; what this means, we understand theologically in the light of Jesus Christ. As creatures of a particular kind, human beings have both proximate and ultimate ends, which are God's gift to us in creating us as beings of this kind. Our good or flourishing consists in the fulfillment of these ends.*

Karl Barth's account of health begins with the knowledge that we are commanded and set free by our Creator "to exist as a living being of this particular, i.e., human structure."[1] To understand health theologically,

---

1. Karl Barth, *Church Dogmatics*, ed. and trans. G. W. Bromiley and T. F. Torrance, 13 vols. (Edinburgh: T&T Clark, 1956-75), III.4, p. 32. Hereafter *CD*.

therefore, we must understand what it means to be a creature of this particular kind. For Barth, we can only know the reality of what this means in and through Jesus Christ.[2] If we attempt to construct a theological anthropology grounded on our own experience or empirical scientific investigations of human nature, we shall encounter two problems. One is that empirical investigations can only disclose the "phenomena" of human existence, not the reality (though this does not mean that scientific insights have no part to play, as we shall see). The other is that our human nature is radically corrupted by sin, so we can never be confident that we are observing human nature as God created it to be, rather than a sinful distortion of that nature. Only in Jesus Christ do we find human existence uncorrupted, as God means it to be. According to Barth, in Jesus Christ we know ourselves to be a kind of creature elected and called to relationship with God and personal responsibility before God; our creaturely being as God's covenant-partners is also "a being in encounter," in "I-Thou" relationships; a human creature is an integrated being, "a substantial organism, the soul of his body"; and finally we know ourselves to be temporal creatures, made to live a finite existence.[3] These four aspects of our creaturely nature correspond to the four dimensions of God's liberating command in Barth's ethics of creation: freedom before God, freedom in fellowship, freedom for life, and freedom in limitation.[4]

To explicate what it means to be creatures of our particular kind, Thomas Aquinas used Aristotelian categories, describing humans as substances of a particular kind, with a particular nature — a particular combination of form and matter. As we saw in chapter 3, this is not a move that commended itself to Barth, who might well have regarded it as an example of the type of "speculative theory of man . . . [which] belongs to the context of a worldview"[5] — in other words, a non-theological "kind of basic discipline which imposes its criterion on all other knowledge and perhaps claims to embrace it."[6] Yet as I suggested in chapter 3, it is possible to read Thomas's use of Aristotle in quite another way: as a *critical* appropriation of certain philosophical tools and concepts, according to theological criteria, in order to do theological work. Eugene Rogers, in the extended conversation he sets up between Thomas and Barth, attempts to bring

2. Barth, *CD* III.2, pp. 19-54.
3. Barth, *CD* III.2, §§ 44-47; quotations at pp. 203, 325.
4. Barth, *CD* III.4, §§ 53-56.
5. Barth, *CD* III.2, p. 22.
6. Barth, *CD* III.2, p. 21.

Thomas as close as possible to Barth at this point, by arguing that in Thomas's thought, nature cannot be understood apart from grace, nor recognized without knowledge of nature's end, which is graciously given by God: "We ought to define nature in terms of grace because it takes Jesus Christ to tell us what nature is."[7] If Rogers is correct to attempt this kind of *rapprochement* between the approaches of Thomas and Barth, the project of developing a theological understanding of creaturely human flourishing informed by both of them has some hope of being a coherent one.

Moreover, Christians need have no difficulty in affirming both a theological understanding of humans as creatures of a particular kind and an evolutionary biological understanding of human origins, because these two accounts are different kinds of narrative addressing different questions.[8] One way of articulating the relationship between them is to say, with Barth, that revelation discloses the reality of human nature, while the natural sciences are concerned with the "phenomena" — those features of the material world that are accessible to human observation and susceptible of theorization in terms of (efficient) cause and effect. Provided it does not acquire the pretensions of a "worldview," this type of scientific inquiry cannot be considered an enemy of a theological anthropology, and can in fact have a positive value:

> Scientific anthropology gives us precise information and relevant data which can be of service in the wider investigation of the nature of man, and can help to build up a technique for dealing with these questions. . . . It is not concerned with his reality, let alone with its philosophical foundation and explanation. But it reveals the plenitude of his possibilities.[9]

---

7. Eugene F. Rogers Jr., *Thomas Aquinas and Karl Barth: Sacred Doctrine and the Natural Knowledge of God* (Notre Dame: University of Notre Dame Press, 1995), p. 190.

8. See further Neil Messer, *Selfish Genes and Christian Ethics: Theological and Ethical Reflections on Evolutionary Biology* (London: SCM, 2007), pp. 50-53.

9. Barth, *CD* III.2, p. 24. I am indebted to my doctoral student, Rev. Philip Chapman, for prompting me to become better acquainted with this aspect of Barth's thought. Perhaps, though, Barth overstates the sharp differentiation he makes between theological and scientific anthropologies (p. 26): when he comes in the ethics of creation to discuss what is meant by health and what "respect for life" requires in this regard (*CD* III.4, pp. 356-375), it becomes necessary to engage with the phenomena of human existence as well as its "true reality and essence" (III.2, p. 26), and therefore to learn from medical science as well as revelation. Furthermore, it would be a mistake to learn from Barth's rejection of anthropological

This differentiation between scientific and theological perspectives on human being suggests a response to one problem for a teleological account of health noted in chapter 2, namely Garret Merriam's rejection of an Aristotelian notion of natural kinds as a form of "species essentialism" that is untenable after Darwin.[10] While others, such as Daniel Sulmasy, do still wish to defend natural kinds,[11] it might be more promising theologically to think of ourselves, and other living things, as members of *creaturely* rather than natural kinds. This will mean locating our talk of the goods, goals, and ends proper to human kind within an understanding of the good purposes of the Creator, rather than trying to read claims about these goods and ends off modern biological accounts that tend methodologically to bracket out considerations of the good.

Our flourishing, as creatures of a particular kind, consists in the fulfillment of the ends proper to that kind of creature. There is a very close relationship between the kind of creature we are and the ends towards which

---

"worldviews" a general hostility to philosophical insights: he is willing enough (for example) to learn from the existentialism of Karl Jaspers about "the historicity of man and his relatedness to another," an insight that he is ready to acknowledge as a new and "decisively important . . . phenomenon of the human" (III.2, p. 113). A third caveat must also be entered in connection with Barth's view of the relationship between scientific and theological anthropologies: his own treatment of human evolution is sketchy and inaccurate (III.2, pp. 79-85). For example, he lumps Lamarck and Darwin together as exponents of the same theory and, bizarrely, presents Ernst Haeckel rather than Darwin as its chief representative (p. 80). He also gives the impression of treating evolutionary theory as hostile to theological anthropology (pp. 89-90). Perhaps in the context of the German-speaking world of the late 1940s, all this is understandable: neo-Darwinian evolution had only relatively recently become firmly established as a biological paradigm through the "modern synthesis" of Darwinian natural selection with Mendelian genetics in the 1920s and 1930s; Barth's engagement with Darwinian theory appears to be mostly mediated through three theological apologists, Otto Zöckler, Rudolf Otto, and Arthur Titus, all of whose knowledge of Darwinism would have predated the modern synthesis. His own direct engagement with biological science in *CD* III.2 is mostly with his non-Darwinian Basel colleague Adolf Portmann (pp. 85-90). And with the Third Reich a very recent memory, it would be very easy to associate Darwinian evolution with the kind of distorted evolutionary ideology — mediated indeed to some extent through Haeckel — that informed the theory and practice of Nazism. For all this, however, Barth does allow that evolutionary biology can give knowledge of human phenomena, something that is much clearer in our time than it was in his.

10. Garret Merriam, "Rehabilitating Aristotle: A Virtue Ethics Approach to Disability and Human Flourishing," in *Philosophical Reflections on Disability*, ed. D. Christopher Ralston and Justin Ho (Dordrecht: Springer, 2010), pp. 133-151 (p. 134).

11. Daniel P. Sulmasy, "Dignity, Disability, Difference, and Rights," in Ralston and Ho, *Philosophical Reflections on Disability*, pp. 183-198.

we are properly directed: in the Aristotelian language that informs Thomas's account of teleology, the form is the final cause. Furthermore, because a Christian doctrine of creation understands the existence of creatures of a particular kind to express the purpose of a good and loving Creator, it is possible to recognize the ends or goals proper to our nature as *good* in an "evaluative," not merely a "functional" sense. As I shall explain more fully under thesis (iii) below, this offers a way around the *impasse* I noted in chapter 1, in relation to the dispute between Christopher Megone and Jerome Wakefield, concerning the relationship between natural function and the good. Finally, as later theses will describe more fully, a teleology informed by the Thomist account discussed in chapter 3 will recognize both an *ultimate* end of human life — union with God — and *proximate* or this-worldly ends proper to our nature: there are things in this present, mortal life that it is good for creatures of our kind to be and to do.

(ii) *Our ultimate end is life in union with God, which means that theological talk of human goods and ends (including health) must be placed within an eschatological horizon; but some time-honored Christian ways of understanding that horizon require critical theological reappraisal.*

According to the Letter to the Ephesians, the divine purpose in creation is "to gather up all things in [Christ], things in heaven and things on earth"(1:10), and in the Thomistic account of teleology surveyed in chapter 3 there is a sense in which all created things, just because they are God's creatures, have God as their ultimate end. However, there is an additional sense in which only humans and other rational creatures can have God as their last end: only rational creatures can know and love God, and so enjoy the beatific vision.[12] According to Thomas, the soul does not need the body to attain this perfect happiness, though when the soul is reunited with the resurrected body, its happiness will increase "not in intensity, but in extent."[13]

Since Thomas acknowledges a sense in which this ultimate end will only be completely fulfilled at the resurrection of the dead, it is clear that in this perspective human teleology must be understood eschatologically. Yet there are various ways in which this Thomistic perspective requires theological reappraisal. One is its understanding of the soul and its relationship to the body, heavily influenced by Aristotelian biology. To the ex-

---

12. Thomas Aquinas, *Summa Theologica* I-II, q. 1, a. 8, responsio. Hereafter *ST*.
13. Aquinas, *ST* I-II, q. 4, a. 5, ad 5.

tent that aspects of the latter are *genuinely* discredited by post-Darwinian biological science, Thomas's understanding of soul and body will require some reconsideration. Dialogue with modern biology has indeed been one of the stimuli of recent work on biblical and theological anthropologies.[14] Space does not permit a full review of this complex area of discussion, but in general it can be said that much of this work tends to emphasize more strongly the significance of the resurrection of the body over against the post-mortem survival of the soul in speaking of the Christian eschatological hope. If such reappraisals are justified, their effect will be to strengthen the present thesis by accentuating the eschatological horizon against which the ends of human life must be understood.

Another respect in which some critical reappraisal of the tradition is in order is the importance of rationality in relation to humanity's ultimate end. For Thomas, reason — albeit understood, under Aristotle's influence, in a very broad way — is of central importance in making it possible for humans to attain their last end. Yet theological reflection on cognitive impairments by authors such as John Swinton and Amos Yong suggests that the experience of disability might call this valorization of rationality into question. For Swinton, for example, this reminds us of the danger that striving for intellectual knowledge *about* God can distract us from receiving the knowledge *of* God and the invitation to participate in the life of God that are offered us as gifts. For Yong, questions about the resurrection life of someone with (for example) Down syndrome demonstrate the need for an expansion of theological imagination in speaking of the eschatological hope promised by God — always remembering that any discussion of our eschatological future is an attempt to delve into the most profound of mysteries, and calls for a good measure of reticence in our speculations. In this life, at best, we see through a glass darkly.

A third, related area of critical reappraisal concerns the persistent ambivalence about the goods of embodied human life that the Christian tra-

---

14. This dialogue must be handled with some care: in the theological perspective articulated in this book, the natural sciences should not be regarded as independent sources of authority to be weighed in the balance against Scripture, but as standing in what Richard Hays calls a "hermeneutical relation" to Scripture: helping us to understand and interpret Scripture in our own context. In relation to the present discussion, for example, insights from the natural sciences can properly challenge partial or distorted readings of Scripture, and perhaps suggest directions for more complete or fruitful readings, in the development of a Christian theological anthropology. See Richard B. Hays, *The Moral Vision of the New Testament: Community, Cross, New Creation* (Edinburgh: T&T Clark, 1997), p. 310.

dition has often learned, rightly or wrongly, from Thomas. In this regard, one valuable resource for critical theological reappraisal is Dietrich Bonhoeffer's account of the ultimate and the penultimate, which gives us a language with which to speak of the goods of embodied life in this world as genuine, albeit finite, goods.[15] They are to be valued, not finally for their own sake, but for the sake of the ultimate — which *both* underwrites their own proper value *and* preserves them from having to bear the burden of our ultimate hopes and fears.

(iii) *There are penultimate or proximate ends that are both objective and universal, given simply by virtue of our being creatures of this particular, human, kind. Knowledge of these ends comes in the first place from God's revelation in Christ, rather than any empirical study of the human species, but many kinds of human knowledge and insight can be critically assimilated to this theological understanding.*

Thomas Aquinas's account of human beings as substances of a particular kind — matter given a particular form by a human soul, and directed to certain connatural ends — uses Aristotelian concepts and language to explicate the theological insight that to be a human creature, in the good purposes of God, just is to be an entity whose good consists in the fulfillment of certain goals and ends rather than others. So some goods, goals, and ends of human life are — we might say — "objectively" given as part of the structure of a creaturely life of this kind, and are common to all human beings *as* human beings. This is a very different understanding from many of the philosophical accounts surveyed in chapter 1, such as that of Lennart Nordenfelt.[16] Just as Nordenfelt acknowledges the tension or contrast between his own "want-satisfaction" view of happiness and an Aristotelian conception of *eudaimonia*,[17] his view similarly contrasts with the theological teleology being developed here, for all its differences from Aristotelian teleology.

---

15. Dietrich Bonhoeffer, *Ethics*, Dietrich Bonhoeffer Works, vol. 6, ed. Ilse Tödt, Heinz Eduard Tödt, Ernst Feil, and Clifford J. Green, trans. Reinhard Krauss, Charles C. West, and Douglas W. Stott (Minneapolis: Fortress, 2005), pp. 146-170.

16. As noted in chapter 1, Nordenfelt does recognize some universal vital goals, which turn out to be those having to do with survival, and does also believe that many other vital goals will be widely shared; but nonetheless, he does hold that apart from a relatively small number of "basic vital goals," all vital goals are subjectively chosen.

17. Lennart Nordenfelt, *Health, Science, and Ordinary Language* (Amsterdam: Rodopi, 2001), pp. 118-119, quoted by Iain Law and Heather Widdows, "Conceptualizing Health: Insights from the Capability Approach," *Health Care Analysis* 16 (2008): 303-314 (p. 306).

The claim that there are "objective" and "universal" penultimate ends given in the structure of human creaturely life requires some important qualification. First, we cannot learn about those ends simply by theoretical analysis or empirical investigation of human life, because the claim is a theological one about God's purposes in creation. In Karl Barth's terms, as noted already in connection with thesis (i), science can only yield knowledge of the "phenomena"; to understand the reality of things as reflective of God's good purposes, we depend on revelation. This means that a confident claim about the objectivity and universality of human goods, goals, and ends should be combined with a good deal of epistemological humility about our capacity to know what those goods, goals, and ends are. Furthermore, our knowledge — even the way we understand the revelation we have received from God — is all too easily distorted by sin in such forms as self-deception and willful ignorance.[18] This theologically-motivated suspicion concerning the damage done to our understanding by our sinfulness should encourage in us a self-critical alertness to the ways in which our concepts of health and human flourishing might be distorted by the ideologies, deceptions, and self-deceptions to which we are vulnerable in the particular cultural contexts we inhabit. It should also, therefore, make us hospitable to insights that can help unmask those ideologies and deceptions, such as the critiques developed in the light of disability.

For Barth, as we have seen, knowledge of the "true reality and essence"[19] of humanity is revealed in Jesus Christ. The goods, goals, and ends given in the structure of human creaturely life, in other words, must be understood first and foremost in the light of Christ. Once that is established, such a Christocentric perspective can be thoroughly hospitable to many diverse insights — including, in the context of this book, biological and psychological understandings of human functioning, philosophical arguments about flourishing, disease, and illness, and reflections on the diverse and complex experience of disability — always provided that such insights are *critically* assimilated into our understanding, according to theological criteria. In relation to the argument between Wakefield and Megone about teleology and the good, discussed in chapter 1, this theological perspective will agree with Wakefield that judgments about the good of living beings (in an "evaluative" sense) cannot be derived from modern biology. But in this perspective, they do not have to be derived from biology: the evalua-

---

18. Cf. Barth, *CD* IV.3.1, pp. 368-478.
19. Barth, *CD* III.2, p. 26.

tion of goals and ends as *good* is a theological judgment, part of an understanding of God's good purposes in creation developed in response to the biblical witness to God's revelation in Christ. When the "evaluative" judgment is made theologically, the way is open for the data and theories of modern biology to make their own proper contribution in filling out this understanding, as (to paraphrase Barth) commentary on a text that has already been "known and read for itself."[20]

(iv) As well as those penultimate goods, goals, and ends common to all human creatures, there are goals and ends that are particular and contextual, aspects of the vocations of particular human lives. Discernment of both these particular vocational goals and the appropriate routes towards universal ends in particular contexts will require habits of attentiveness to God's command, or, put another way, the virtue of practical wisdom or prudence.

There are many different routes through this life toward our ultimate end, and we must pick our way through the thicket of diverse and sometimes conflicting penultimate goods with which this life presents us. Even universally shared human ends (proximate as well as ultimate) must be pursued in the particular circumstances of individual lives, so each of us needs to discern the best way to do so. Moreover, there are goals that are not universally human, but particular to the diverse vocations that each of us is called to pursue. Because our creaturely existence is temporal and finite, lived in particular circumstances and with particular gifts and abilities, the human life that God calls us to live will always embody a particular vocation; each person's obedience will take a particular form. As Barth puts it, "[t]he fact that [God] has a special and constantly renewed intention for each means that the freedom of obedience to which He calls man will always be a freedom in limitation."[21]

As we saw in chapter 3, a Thomist teleology suggests the importance of the virtue of prudence in enabling us to choose the best path through the various proximate goods open to us.[22] As I emphasized against Garret Merriam in chapter 2, this understanding of the role of prudence or practical wisdom does not support an entirely individualist or relativist notion of human flourishing. Prudence is better understood as the virtue that en-

---

20. Barth, *CD* III.2, p. 122.

21. Barth, *CD* III.4, p. 595. For his full discussion of vocation in the context of "freedom in limitation," see pp. 595-647.

22. See Aquinas, *ST* I-II, q. 10, a. 2; II-II, q. 47, a. 7.

ables us to discern how the goods, goals, and ends of human life (some of which, as I have already remarked, will be both objective and universally valid) are best instantiated in the specific circumstances of *this* particular human life. In Jean Porter's words,

> Prudence, which takes account of the specifics of an individual's own character and circumstances, determines what, concretely, it means for this individual to be in accordance with reason.... That is to say, prudence determines what amounts to a substantive theory of the human good, at least as it applies to this individual in his particular setting, although of course the individual may not be able to formulate that theory in any systematic way.[23]

Such an emphasis on prudence is sometimes thought to sit uneasily with Barth's robustly theological divine-command ethic of creation. For example, Brian Brock has taxed me with using prudence, in an earlier and far less developed account of health and disease, as a theologically impoverished "middle axiom" to mediate between a biblically-rooted understanding of God's salvific activity and concrete, practical moral judgments.[24] However, I do not think we can do without prudence, or something very like it, in this field. Nor need it be reduced to an impoverished middle axiom (however unsuccessful my own earlier account might have been at avoiding that danger). If, as Barth says, God "has a special and constantly renewed intention for each,"[25] then it seems likely that some gift of discernment, or habit of alertness and attentiveness to God's command, must be part of that growing character which he calls "a work of the grace of God on man ... the distinctive molding and determining of the course and form of life from the center of the Thou as which he is addressed by God."[26] Moreover, it is widely agreed

---

23. Jean Porter, *The Recovery of Virtue: The Relevance of Aquinas for Christian Ethics* (London: SPCK, 1994), p. 162.

24. Brian Brock, Review of Celia Deane-Drummond, ed., *Brave New World: Theology, Ethics, and the Human Genome Project* (London: T&T Clark, 2003) and Celia Deane-Drummond, Bronislaw Szerszynski, and Robin Grove-White, eds., *Re-Ordering Nature: Theology, Society, and the New Genetics* (London: T&T Clark, 2003), *Studies in Christian Ethics* 19.1 (2006): 110-116 (pp. 113-114); Brian Brock, "Why the Estates? Hans Ulrich's Recovery of an Unpopular Notion," *Studies in Christian Ethics* 20.2 (2007): 179-202 (p. 188). The account criticized by him is Neil Messer, "Health, the Human Genome Project, and the 'Tyranny of Normality,'" in Deane-Drummond, ed., *Brave New World*, pp. 91-115.

25. Barth, *CD* III.4, p. 595.

26. Barth, *CD* III.4, pp. 388-89.

that on the whole Barth does not regard the command of God as arbitrary, but as a gracious summons to conformity with that structure of creaturely existence that is God's good gift in creation.[27] That being the case, it is possible to speak of a form of Christian wisdom — itself the work of God's sanctifying grace — which enables the individual to discern what that summons entails in the particular circumstances of his or her life. We might call that wisdom "prudence"; and we might notice that such an understanding of prudence, developed from Barth's ethic of the command of the Creator, looks not unlike Jean Porter's Thomist account quoted above.

A further important emphasis concerning the virtue of prudence, so understood, has long been familiar from the work of Stanley Hauerwas and others: the formation of Christian virtues and character depends on participation in the Church's life and distinctive practices. Such participation brings about (or ought to bring about) a transformation of moral vision, as he and William Willimon remark in the context of a story about disability.[28]

Health and Creaturely Flourishing

(v) *Health is not to be equated with the whole of human well-being (as in the World Health Organization definition), and Christian talk of "wholeness" should be treated with considerable caution. Nonetheless, there are important insights expressed in talk of wholeness that can and should be articulated in other ways.*

The dangers of the WHO definition of health ("a state of complete physical, mental and social well-being . . .") were discussed in chapter 1: it claims too much territory for "health" (and by extension, for health care and health professionals), while giving too limited an account of human well-being. While many Christian authors have connected the WHO definition with the Hebrew term *shalom* and with a concept of "health as wholeness,"[29] this connection should be treated with considerable caution.

---

27. Nigel Biggar, *The Hastening That Waits: Karl Barth's Ethics* (Oxford: Clarendon, 1993), p. 49.

28. Stanley Hauerwas and William H. Willimon, *Resident Aliens* (Nashville: Abingdon, 1989), p. 95. The context of this remark is the story of Buncombe Street Church and Dorothy, quoted and discussed above in chapter 3 (pp. 151-152).

29. E.g., John Wilkinson, *The Bible and Healing: A Medical and Theological Commentary* (Grand Rapids: Eerdmans, 1998), pp. 11-13; Abigail Rian Evans, *Redeeming Marketplace Medicine: A Theology of Health Care* (Eugene, OR: Wipf and Stock, 2008).

It could lead Christians to operate with too restricted a notion of human creaturely flourishing, and obscure or distort our understanding of the relationship between health and other goods of embodied creaturely life. Furthermore, since "wholeness" is an eschatological promise, conceptualizing health as wholeness could confuse and obscure our understanding of the relationship between penultimate and ultimate goods, goals, and ends.

A more expansive picture of human flourishing than the WHO definition is suggested by Karl Barth's ethics of creation, which identifies the four dimensions to the liberating command of the Creator cited earlier: freedom before God, freedom in fellowship, freedom for life, and freedom in limitation.[30] This wide-ranging picture articulates one of the insights that Christians try to express through the language of wholeness: that human *flourishing* or *well-being* must indeed be understood comprehensively, to encompass every aspect of creaturely human existence. There are human goods, goals, and ends given by God in creation that have to do with our relationship with God, our human relationships and social structures, and the embodied lives we lead — in each of these respects, there are ways of living that fulfill God's good purposes for human creatures. Barth adds the fourth dimension of "freedom in limitation," as noted above under thesis (iv).

The boundaries between these different aspects of the command of the Creator are clearly not watertight, because they are all aspects of what it means to flourish as this kind of creature. So, for example, a loss of the "strength for life" (as Barth defines health) could profoundly affect a person's human relationships in many different ways, and, conversely, health is often affected by relationships, social structures, economic conditions, and so forth.[31] This is another of the insights that Christians seek to express through wholeness-talk. However, this does not mean that "freedom in fellowship" can simply be collapsed into "health," or that there is no distinction between "freedom in fellowship" and "freedom for life." The different aspects of the command of the Creator can and should be distinguished from one another, even if they cannot be separated.

(vi) *Karl Barth's definition of health as "strength for human life"*[32] *expresses some themes that are at the very heart of a theological understanding of health: it recognizes the character of embodied human life as a divine gift that*

---

30. Barth, *CD* III.4, §§ 53-56.
31. Barth, *CD* III.4, p. 363.
32. Barth, *CD* III.4, p. 356.

comes with a divine call, it offers an account of physical and mental integration that captures another important insight expressed in Christian talk of wholeness, and it allows a balanced understanding of health as the "capability, vigor, and freedom" to live a human life.

Our embodied creaturely life is both gift and summons. It is the good gift of God the Creator, and with the gift comes the Creator's liberating command of "respect for life."[33] One aspect of that command is to "will to be healthy":[34] to be receptive to God's gift of this particular kind of creaturely flourishing. According to Barth, the command to "will to be healthy" underpins a Christian mandate for medicine and health care, while emphasizing that the strength for life — the ability to answer God's summons and live a flourishing human life — is itself God's gracious gift.

Barth expands his definition in two ways that offer important connections with the discussions earlier in this book. One is to describe it as "the integration of the organs for the exercise of psycho-physical functions."[35] As observed in chapter 3, he thinks of the will to be healthy as the will to be a whole, integrated human being. This picture of physical and psychological integration can give an account of another of the important insights expressed in Christian talk of wholeness, without being prone to the distortions discussed under thesis (v).

The other is to describe the strength for life as "capability, vigor, and freedom" — or to borrow a phrase from Amartya Sen, the capacity for all the "doings and beings" that constitute an embodied life of this particular, human, kind.[36] Critically appropriated, the capability approach to health outlined by Iain Law and Heather Widdows, drawing on the work of Sen and Martha Nussbaum, could help to "flesh out" (almost literally) what might be meant by the "capability, vigor, and freedom" needed to live a human life.[37]

The appropriation of insights from this capability account must be critical in at least two different ways, which might appear to be in tension with one another. The first concerns what Sen calls the "incompleteness"

---

33. Barth, *CD* III.4, § 55.1.
34. Barth, *CD* III.4, p. 357.
35. Barth, *CD* III.4, p. 356.
36. Amartya Sen, "Capability and Well-Being," in *The Quality of Life*, ed. Martha C. Nussbaum and Amartya Sen (Oxford: Clarendon, 1993), pp. 30-53 (p. 31).
37. Law and Widdows, "Conceptualizing Health." I am grateful to Professor Kenneth Vaux for first suggesting to me that the capability approach could be theologically fruitful for an understanding of health.

of his capability approach. As we saw in chapter 1, he refrains from giving a determinate account of the "doings and beings" that make for a flourishing human life, partly to enable the capability approach to be used by those holding a range of moral theories with rival visions of the human good.[38] Though Law and Widdows hold the incompleteness of his approach to be an advantage in conceptualizing health, Martha Nussbaum has criticized Sen on this point, arguing that the capability approach would benefit from the more determinate, objective account of human goods offered by an Aristotelian account of human flourishing.[39]

Law and Widdows's support for the incompleteness of Sen's capability approach signals the importance of a proper recognition of the contextual character of human life and needs. However, in the theological perspective being developed here, there must be a limit to the contextual variability of our understanding of health. The theological teleology outlined in chapter 3 suggests that there are *givens:* "doings and beings" that must always be part of what it means for a creaturely life of this kind to flourish. Thus far I side with Nussbaum against Sen on the question of incompleteness.

Only thus far, however, because the other critical perspective that needs to be brought into play is connected to the epistemological humility required by my theological account. We see through a glass darkly, and our vision is very easily clouded by the distorting effect of our sinfulness on our understanding, will, and desire. So in the context of a discussion of health, we need to be keenly alert to ways in which we might have come to understand human flourishing in partial or distorted ways, particularly when those distorted understandings of what it means to flourish serve to exclude or marginalize some of our neighbors. This theological perspective will therefore be open to the critique of the concept of "normal function" in the context of disability developed by Ron Amundson, Anita Silvers, and others.[40] It will also be receptive to the argument of theologians

---

38. Sen, "Capability and Well-Being," pp. 48-49.

39. Martha Nussbaum, "Nature, Function, and Capability: Aristotle on Political Distribution," *Oxford Studies in Ancient Philosophy,* supp. vol. (1988); see also Martha Nussbaum, "Non-Relative Virtues: An Aristotelian Approach," in Nussbaum and Sen, eds., *The Quality of Life,* pp. 242-276.

40. Ron Amundson, "Against Normal Function," *Studies in History and Philosophy of Biological and Biomedical Sciences* 31.1 (2000): 33-53; Anita Silvers, "A Fatal Attraction to Normalizing: Treating Disabilities as Deviations from 'Species-Typical' Functioning," in *Enhancing Human Traits: Ethical and Social Implications,* ed. Erik Parens (Washington, DC: Georgetown University Press, 1998), pp. 95-123.

such as John Swinton: that the lives of disabled people with profound and complex needs might direct our attention, both to distorted perceptions of the human good in our cultural context, and to neglected insights about human flourishing from the Christian tradition.[41]

The tension between these two critical themes emerged clearly in chapter 2 in the discussion of Martha Nussbaum's treatment of disability. Nussbaum, it will be recalled, argues that a threshold level of each of her ten basic capabilities is a requirement of "a life that is worthy of the dignity of a human being," and that in the most extreme cases, "[o]nly sentiment leads us to call the person in a persistent vegetative condition, or an anencephalic child, human."[42] In chapter 2 I suggested that a Christian theological ethic ought to object to a conception of human flourishing that would lead us to exclude some offspring of human parents from the status of "human."

The following points may be made towards a resolution of this tension:

(1) There are determinate things that it means for human beings to flourish in this life; so far I agree with Nussbaum.

(2) However, we need to maintain a level of epistemological humility about the partial and provisional character (at best) of our attempts to stipulate *what* flourishing consists in.

(3) Furthermore, we should be open to the possibility, canvassed in chapter 2 in the discussion of Amundson and Wakefield, that there are diverse ways for humans to flourish — that is, there are diverse ways in which particular penultimate ends might be fulfilled. One concrete example is Amundson's argument that spoken and signed language can equally be understood as ways in which the goals associated with language and communication can be realized.[43]

(4) This perspective does allow the possibility that some impairments will genuinely inhibit human flourishing, perhaps severely and (in this life) permanently.

(5) However, in the case of any particular impairment we should not be too ready to leap to the conclusion that it does truly or seriously inhibit flourishing. One reason for caution is highlighted by discussions (by

---

41. John Swinton, "Known by God," in *The Paradox of Disability: Responses to Jean Vanier and the L'Arche Community from Theology and the Sciences*, ed. Hans S. Reinders (Grand Rapids: Eerdmans, 2010), pp. 140-153.

42. Martha C. Nussbaum, *Frontiers of Justice: Disability, Nationality, Species Membership* (Cambridge, MA: Harvard University Press, 2006), pp. 180, 187.

43. Amundson, "Against Normal Function," pp. 42-43.

Amundson, Ian Basnett, and others) of the disparity between first-person and third-person estimates of the effect of impairments on quality of life.[44] Another is that in the theological perspective being articulated here, creaturely flourishing is not in any case coterminous with quality of life.

(6) We should also make a clear distinction between the judgment that certain impairments might severely and permanently inhibit human flourishing in this life and the claim that in the most extreme cases, individuals with those impairments cannot be considered "human" (as Nussbaum argues). These are separate judgments, and there are theological reasons for thinking that human beings are not entitled to make the latter in ways that involve excluding some offspring of human parents from the moral status that we customarily recognize in one another.[45]

(7) Where impairments do severely and permanently inhibit their bearers' flourishing in this life, it becomes all the more important to emphasize the eschatological horizon of a theological account of health, a theme that will be developed further in later theses.

(vii) *A Christocentric account of creaturely flourishing can be hospitable to evolutionary and other biological insights about natural functions, though these insights, like those from other sources, must be* critically *appropriated.*

As I observed at the end of chapter 1, since the concept of health is concerned with biologically embodied human life, evolutionary and other biological insights into the functioning of human organisms can be expected to be particularly important in "fleshing out" what it means to flourish as an embodied human creature. One of the most sophisticated and fully developed evolutionary treatments of health is Jerome Wakefield's account, discussed in chapter 1. Wakefield, it will be recalled, defines disorders (including diseases and injuries) as "harmful dysfunctions," where a dysfunction is "the failure of an internal mechanism to perform a natural function for which it was designed [by natural selection]."[46]

---

44. Ron Amundson, "Quality of Life, Disability, and Hedonic Psychology," *Journal for the Theory of Social Behaviour* 40.4 (2010): 374-392; Ian Basnett, "Health Care Professionals and Their Attitudes Toward and Decisions Affecting Disabled People," in *Handbook of Disability Studies,* ed. Gary L. Albrecht, Katherine D. Seelman, and Michael Bury (Thousand Oaks, CA: Sage, 2001), pp. 450-467.

45. See further Neil Messer, *Respecting Life: Theology and Bioethics* (London: SCM, 2011), pp. 114-118.

46. Jerome C. Wakefield, "The Concept of Mental Disorder: On the Boundary between Biological Facts and Social Values," *American Psychologist* 47.3 (1992): 373-388 (p. 374).

While Wakefield's account can certainly help to inform a theological understanding of health, it must like other insights from any source be *critically* appropriated. Some of the critical caveats to be entered will have to do with the scientific status of claims about natural functions. As Wakefield freely acknowledges, claims about the evolutionary origins of particular natural functions are often highly under-determined by scientific evidence: it would be impossibly rigorous to require an *actual*, well-supported evolutionary explanation for the existence of a structure or mechanism in order for its failure to be recognized as a dysfunction.[47] Speculative claims that some physical or mental mechanism is a natural function with particular evolutionary origins can certainly be heuristically useful in understanding disorders, but the provisionality and tentativeness of such claims must always be kept in view. Related to this is the philosophical critique developed by authors such as Amundson of the ways in which biological evidence is interpreted and used (sometimes, it is argued, in ways distorted by prejudices and ideologies) to support particular notions of "normality," as noted under thesis (vi).

A further critique concerns the limits of this kind of evolutionary reasoning by virtue of its character as a modern scientific enterprise. In particular, it is only concerned with efficient-causal explanation and makes a sharp separation of "fact" from "value." This means, as became clear in Christopher Megone's dispute with Wakefield surveyed in chapter 1, that modern evolutionary biology cannot supply a properly teleological understanding of health; but neither is it competent to refute such an understanding, since modern biology methodologically brackets out precisely the questions of final causation and "value" that would be needed to develop such a teleological account. In short, neo-Darwinian biology cannot comment either way on the validity of the kind of Aristotelian teleology advocated by Megone.

If we are developing a *theological* teleological account of health, this limitation is emphasized all the more by the theological understanding of the role and limits of natural science noted under thesis (i) above. In Barth's terms, the natural sciences are only competent to deal with the "phenomena" of the world, and their use theologically is as "commentary on a text which must first be known and read for itself."[48] That commentary can be enormously valuable within its limits, but we must keep those

---

47. Wakefield, "The Concept of Mental Disorder," pp. 382-383.
48. Barth, *CD* III.2, p. 122.

limits clearly in mind. We shall not learn from biology that human beings are *creatures* of a particular kind, but neither is biology competent to deny this (whatever "New Atheist" authors might think). However, if we have learned from a Christian doctrine of creation that we *are* creatures of a particular kind, biology can tell us a great deal about the features and conditions of this kind of creaturely life. Again, we cannot expect biology to identify the proper ends of this kind of creaturely life, though it might have valuable "commentary" to add to an understanding of human creaturely ends that we gain from theological sources.[49]

(viii) *Health is a* penultimate *good, concerned with the fulfillment of proximate human goals or ends, although it points toward the ultimate fulfillment of human life. If it is treated as an ultimate end — which is a form of idolatry — the results are likely to prove destructive.*

As I suggested above (thesis ii) and in chapter 3, health is — to use Dietrich Bonhoeffer's language — a penultimate, not an ultimate good. That is to say that in terms of the Thomist account of teleology explored in chapter 3, health is concerned with the fulfillment of some of the *proximate* ends of embodied creaturely life. To be sure, it points beyond itself to the ultimate fulfillment in eternity of the real but finite goods of creaturely life in this world. There may be respects in which eternal life will complete goals represented in this life by health, as is hinted (for example) in the closing chapters of the book of Revelation, in John's vision of the heavenly city in which "[d]eath will be no more; mourning and crying and pain will be no more" (Rev. 21:4), and the leaves of the tree of life will be "for the healing of the nations" (Rev. 22:2). But that ultimate fulfillment by God's grace represents the transformation, not merely the continuation, of the goods of this-worldly health; and such transformation cannot be delivered by any human power or agency, including medical agency. In the present age, as Barth remarks, "health, like life in general, is not an eternal but a temporal and therefore a limited possession."[50] The indefinite extension of this present human life, as envisaged in some transhumanist and "radical

---

49. This point is often misunderstood in discussions of evolution and ethics: see further Messer, *Selfish Genes*, pp. 116-121, and Stephen J. Pope, *Human Evolution and Christian Ethics* (Cambridge: Cambridge University Press, 2007), pp. 291-293; *contra* Larry Arnhart, "The Darwinian Moral Sense and Biblical Religion," in *Evolution and Ethics: Human Morality in Biological and Religious Perspective*, ed. Philip Clayton and Jeffrey Schloss (Grand Rapids: Eerdmans, 2004), pp. 204-220.

50. Barth, *CD* III.4, p. 371.

life extension" scenarios,[51] would not be its transformation or fulfillment. It would more likely represent the corruption of the genuine goods of health and mortal life, by treating them as matters of ultimate importance — that is, idolizing them or investing in them the devotion only proper to God. As Colin Gunton once remarked, "If the created order, or part of it, is treated as god, then it behaves like god for those who so treat it, but for destructive rather than creative ends."[52] It is not only extreme, speculative projects like radical life extension, however, that are prone to this dynamic: the pursuit of health by more mundane and unspectacular means can likewise become idolatrous if it is elevated to an ultimate goal.

(ix) *Health and the "will to be healthy" are not merely individual, but also social and political matters.*

As we saw in chapter 3, Barth puts this very clearly and sharply, emphasizing the importance of social conditions in affecting health for good or ill, and contemplating the possibility that radical societal change might be needed to remove social obstacles to the health of some individuals, groups, or communities. "Where some are necessarily ill," he remarks, "the others cannot with good conscience will to be well."[53] Many of the social, economic, and political determinants of health, for good or ill, are by now well known.[54] As Barth suggests, they include forms of socioeconomic relationship that give rise to poor pay, long working hours, unhealthy or dangerous working conditions, and poor housing. More generally, eco-

---

51. An approach to the indefinite postponement of aging referred to as "strategies for engineered negligible senescence" (SENS) is promoted, for example, by Aubrey de Grey and the SENS Foundation, though the latter resists the label "transhumanist": see its website at http://www.sens.org. For presentations of the SENS approach, see Aubrey D. N. J. de Grey, Bruce N. Ames, Julie K. Andersen, Andrzej Bartke, Judith Campisi, Christopher B. Heward, Roger J. M. McCarter, and Gregory Stock, "Time to Talk SENS: Critiquing the Immutability of Human Aging," *Annals of the New York Academy of Sciences* 959 (2002): 452-462, and Aubrey de Grey with Michael Rae, *Ending Aging* (New York: St. Martin's, 2008).

52. Colin Gunton, *The Actuality of Atonement: A Study of Metaphor, Rationality, and the Christian Tradition* (Edinburgh: T&T Clark, 1988), p. 72. For a range of critical theological perspectives on such scenarios, see Celia Deane-Drummond and Peter Manley Scott, eds., *Future Perfect? God, Medicine, and Human Identity* (London: T&T Clark, 2006), and Brent Waters, *From Human to Posthuman: Christian Theology and Technology in a Postmodern World* (Aldershot: Ashgate, 2006).

53. Barth, *CD* III.4, p. 363.

54. See, e.g., Michael Marmot and Richard G. Wilkinson, eds., *Social Determinants of Health*, 2nd ed. (Oxford: Oxford University Press, 2006).

nomic and political conditions that limit people's access to clean water, adequate food, and other basic necessities have a massive impact on health globally, and access to health care is notoriously uneven both within and between nations.[55] A third area of concern suggested by Barth's remark is the unequal access to medicine and health care between rich and poor, both within any one nation and globally. But in addition to these quite familiar sociopolitical aspects of health, the effect of social, political, and cultural contexts on the *understanding* of health — for good or ill — must also be remembered. As Jürgen Moltmann remarked some years ago, understandings of what counts as "healthy" vary significantly over time in response to the values and demands of different societies, "[b]ut this does not mean that these ideas of 'health' are necessarily healthy in themselves."[56] Some could be alienating or dangerously false: one need only think of the toxic notions of health embodied in the early twentieth-century eugenics movement, or the labeling of political dissidents as mentally ill in the Soviet Union, to appreciate the force of Moltmann's point. But his warning against "unhealthy" or distorted concepts of health could also raise more current and less sharply-drawn issues about what counts as "healthy." It could, for example, direct attention back to the critique of disability activists and scholars, discussed above under thesis (vi), that contemporary understandings of health are informed by discourses of normality which misrepresent some forms of physical or psychological difference as disorder. The complex relationship between disease and disability will be explored further below under thesis (xii).

In chapter 1, I suggested that one strength of the WHO definition of health is that it draws attention to the social, political, and economic dimensions of health. However, the WHO definition tends to put matters backwards, claiming that health is necessary for good social and political relations, and at least implying that social and political matters would run a good deal better if doctors had more say in them. The connections made in this thesis, in response to Barth's comments on the social dimension of the will to be healthy, suggest rather that the quality of social relationships

---

55. See further Louis J. Currat, "The Global Health Situation in the 21st Century: Aspects from the Global Forum on Health Research and the World Health Organization in Geneva," *International Review of Mission* 95.376/377 (2006): 7-20; Christina de Vries, "Why Do Churches Need to Continue to Struggle for Health for All?" *International Review of Mission* 95.376/377 (2006): 21-35.

56. Jürgen Moltmann, *God in Creation: An Ecological Doctrine of Creation*, trans. Margaret Kohl (London: SCM, 1985), pp. 270-271.

and political and economic structures have a profound effect on health, and are in some measure the concern of everyone who has any level of responsibility for the life of the political community.

Disease, Suffering, Evil, and Sin

(x) *Diseases are to be understood as a particular class of threats to creaturely flourishing: internal states or processes that tend to disrupt or threaten the fulfillment of the proximate ends of embodied creaturely human life. However, discrimination between diseases and non-diseases cannot be entirely a matter of clear definition, and will require the exercise of practical wisdom.*

Having defined health as the "strength for human life," Barth goes on to describe disease as "the weakness which opposes this strength."[57] This description can be filled out a little by thinking of diseases and other disorders as threats to the flourishing of embodied creaturely human life. Of course, not every kind of threat to embodied creaturely flourishing is a disease or disorder. However, if health is a "capability," in Law and Widdows's terms, then internal states that tend to hinder the "doings and beings" which contribute to the fulfillment of proximate human ends can be recognized as diseases. If health has to do with the integration of physical and psychological functions for the fulfillment of those ends, then an internal state or process which tends to disrupt that integration, or hinders some structure or function from playing its part in the fulfillment of those ends, can be considered a disease. Some threats and hindrances, of course, are only minor or transient: common colds do hinder creaturely flourishing, but usually not seriously or for long. Furthermore, as several of the philosophical accounts surveyed in chapter 1 suggest, the criterion for recognition of a disease cannot be that it actually does hinder creaturely flourishing: it might, for example, be arrested either by medical intervention or some other means before it can do any actual harm. But with qualifications such as these, if we want at least a rough-and-ready definition of disease informed by the range of theological sources and engagements presented earlier in the book, we might say this: An internal state, condition, or process, which tends to disrupt a physical or mental function such that the fulfillment of a proximate end of embodied creaturely human life is hindered or threatened, is a *prima facie* candidate to be considered a disease.

57. Barth, *CD* III.4, p. 366.

Jerome Wakefield's "harmful dysfunction" model of diseases and disorders might help us in recognizing diseases,[58] provided we keep in mind certain caveats and qualifications that have already been suggested: first, in the perspective of the theological teleology developed in the last and present chapters, function and dysfunction are not value-free factual terms, as Wakefield holds. Secondly, as Ron Amundson argues, there might be diverse ways of achieving biological goals, so some caution is needed about identifying the statistically abnormal with the dysfunctional. Thirdly, this account is not concerned with natural functions (strictly speaking), but with creaturely functioning and flourishing. In the last analysis biology cannot definitively tell us whether something is a creaturely function, in the sense of a function that contributes to the fulfillment of a proper creaturely end — though biology can certainly help a good deal with identifying and understanding such functions.

This account does not supply a clear-cut definition that will enable us to separate out diseases from non-diseases unambiguously. However, not even the most apparently clear-cut philosophical accounts, such as Wakefield's, can do that — as Wakefield freely acknowledges.[59] It is not necessary to adopt wholesale the "Roschian" concept of disease proposed by Scott Lilienfeld and Lori Marino in opposition to Wakefield (chapter 1)[60] to accept that the category of disease inevitably has "fuzzy boundaries," and that in recognizing diseases we rely on what Lennart Nordenfelt calls a "conceptual torso" of unproblematic cases (or Lilienfeld and Marino's "prototypes").[61]

If a theological definition of disease cannot supply clear-cut unambiguous criteria for discerning in all cases what is, or is not, a disease, it seems more promising to suggest that this discernment will require the virtue of prudence (or, as I suggested under thesis (iv), the habit of attentiveness to

---

58. Wakefield, "The Concept of Mental Disorder."

59. Jerome C. Wakefield, "Evolutionary versus Prototype Analyses of the Concept of Disorder," *Journal of Abnormal Psychology* 108.3 (1999): 374-399 (p. 379).

60. Scott O. Lilienfeld and Lori Marino, "Mental Disorder as a Roschian Concept: A Critique of Wakefield's 'Harmful Dysfunction' Analysis," *Journal of Abnormal Psychology* 104 (1995): 411-420. More specifically, it is neither necessary nor desirable to accept their claim that disease is a "pure value" concept, since this presupposes the same problematic fact-value dichotomy as Wakefield, which I have already criticized on theological as well as philosophical grounds.

61. Lennart Nordenfelt, "Establishing a Middle-Range Position in the Theory of Health: A Reply to My Critics," *Medicine, Health Care, and Philosophy* 10 (2007): 29-32 (p. 31).

God's command — a quality that can itself only be God's gracious gift to us). Such prudence or attentiveness is needed to discern what constitutes human flourishing and what threatens it — or in Barth's terms, what "the will to be healthy" requires, what particular forms of "weakness" oppose the "strength for life" and what should be done to resist them — in the specific circumstances of a particular human life. Moreover, I also noted under thesis (iv) Hauerwas's emphasis on the transformation of moral vision brought about by participation in the life of the Church. This transformation of moral vision might well lead Christians to surprising and counterintuitive conclusions about health and disease, as is illustrated by Hauerwas's theological perspective on disability. In his and Willimon's story of Dorothy, for example, just such a distinctively Christian exercise of prudence led Buncombe Street Church to regard the life of a woman with Down syndrome very differently from the way the surrounding culture regarded such lives.[62]

(xi) *There is a complex but important set of connections to be made between disease, evil, and sin.*

Barth's account of disease as the "weakness" opposed to the strength for human life allows him to recognize it unequivocally as evil: an aspect of the chaos or "nothingness" *(das Nichtige)* threatening God's good creation. His understanding of "nothingness" was briefly outlined in chapter 3: it is all that God *did not will* in willing the creation; as such, it has a strange paradoxical existence as the "kingdom on God's left hand." It is not beyond the reach of God's sovereignty, and serves God's purposes despite itself (so to say). It is both creation's rebellion against its Creator and the divine judgment on that rebellion. As we have also seen, Barth holds this understanding of disease as an aspect of "nothingness" in dialectical tension with a contrasting point, that disease can direct our attention beyond the limits of this life to God's eternal goodness; but he insists that this does not in any way lessen the first view of it.

Sin is one aspect (though only one) of "nothingness," and this draws attention to the real but complex relationship between sin and disease signaled already by the New Testament texts on healing, as we saw in chapter 3. For example, in the story of the paralyzed man's healing (Mark 2:1-12) Jesus tells the man, "Your sins are forgiven," and James's instructions about prayer and anointing with oil (James 5:13-18) also seem to as-

---

62. Hauerwas and Willimon, *Resident Aliens*, pp. 93-97.

sociate prayer for healing quite closely with the forgiveness of sins. Yet John's account of the healing of the man born blind (9:2-5) rejects any simple causal connection between sin and the man's blindness. This complex association between sin and disease continues to be reflected in later Christian tradition.

It often seems difficult and unpalatable, both theologically and pastorally, to associate sin with disease. As the texts already cited indicate, we cannot straightforwardly generalize by saying that diseases are punishments for individuals' sins, because innocent people often suffer while guilty people do not — as was already well known to the authors of the Psalms and the book of Job. Ethically and pastorally, it can seem problematic even to allow that diseases are *sometimes* caused by sins, for fear of encouraging a censorious attitude that blames the victims of disease and presumes they deserve what they get. Furthermore, if it could be shown that some patients' diseases were caused by their sins, would those patients find themselves given lower priority for treatment or care? Yet some of these reservations depend on misunderstandings of what the Christian tradition, properly speaking, means by sin.

Although it is often used as a synonym for moral transgression, Alistair McFadyen has argued that the language of sin must be understood first and foremost as theological, not merely moral; its most important function is not the tracking of individuals' moral accountability for the proper allocation of praise, blame, reward, and punishment.[63] Furthermore, sin should not be understood first and foremost as a collection of isolated wrong acts, each individually and freely chosen. It is better understood as a condition that each of us inherits and to which we also contribute by our own willing and choosing, a condition that has individual, social, and structural aspects. As McFadyen argues, it is first and foremost a condition of alienation from God that could be called *idolatry*, which also has its destructive and distorting effects on our relations with one another, ourselves, and the world we inhabit.[64] Or, put another way, to be sinners is to be diverted from our "last end" and ultimate good, as a consequence of which the fulfillment of our penultimate goods, goals, and ends is also disrupted and threatened. As such, it is no surprise that sin is complexly connected to disease in all kinds of ways, but this connection is certainly not a

---

63. Alistair I. McFadyen, *Bound to Sin: Abuse, Holocaust, and the Doctrine of Sin* (Cambridge: Cambridge University Press, 2000), pp. 19-22.

64. McFadyen, *Bound to Sin*, chs. 9, 10.

simple causal connection in which diseases are straightforwardly the results of, or punishments for, sins.[65]

However, even if we cannot neatly allocate blame to individuals for their diseases, might we not be driven to say more generally that disease is God's judgment on human sin, and therefore resistance to it is both futile and impious — we must just humbly take whatever comes our way? Barth considers this possibility, but robustly rejects it as "defeatist thinking, and not at all Christian": it overlooks not only God's continuing command to will to be healthy, but also that "He Himself has already marched against that realm on the left, and that He has overcome and bound its forces and therefore those of sickness in Jesus Christ and His sacrifice, by which the destroyer was himself brought to destruction."[66] This is one thing that we learn clearly from the healing narratives in the Gospels, particularly those that portray Jesus in conflict with the diseases that he heals; and it is this that gives humans the mandate to resist disease by medical and other means, a resistance which might indeed otherwise seem presumptuous and futile. It also, incidentally, suggests a Christian answer to any suggestion that those whose diseases are caused by their sins are less deserving of treatment and care than others: if God has acted in Christ to heal our greatest affliction, and made that healing love available indiscriminately to all, though we in no way deserve it, then it would seem a strange Christian response to suggest that the human work of care and healing should be apportioned more selectively. As Paul Ramsey remarked in one of the first modern theological treatments of health care resource allocation, rather than using criteria of merit or desert to allocate scarce medical treatments, "Men should . . . 'play God' in the correct way: he makes his sun rise upon the good and the evil and sends rain upon the just and the unjust alike."[67]

With that ground cleared, the complex connections between sin and disease can be acknowledged.[68] It is not hard to think of ways in which our sins or sinfulness might threaten our health: for example, when gluttony

---

65. See further George Khushf, "Illness, the Problem of Evil, and the Analogical Structure of Healing: On the Difference Christianity Makes in Bioethics," *Christian Bioethics* 1.1 (1995): 102-120.

66. Barth, *CD* III.4, pp. 367-368.

67. Paul Ramsey, *The Patient as Person: Explorations in Medical Ethics* (New Haven: Yale University Press, 1970), p. 256.

68. For a range of theological perspectives, see further Corinna Delkeskamp-Hayes, ed., Theme Issue on Sin and Disease, *Christian Bioethics* 12.2 (2006).

leads to obesity and the ill health it brings in its train,[69] when drug or alcohol misuse bring about the diseases often associated with them, or when excessive career ambition leads to the ill health associated with overwork and stress. The examples could easily be multiplied, yet even in the most apparently straightforward examples, the picture is hardly ever simple: a detached observer cannot know with any confidence how someone's drug or alcohol use, their overweening ambition, or even their excessive eating might be connected to their own personal history and suffering, or to the wrongs done to them by others. All of this reflects the pervasiveness and complexity of the sinful condition in which we find ourselves, as an account such as McFadyen's makes clear. And if many examples can be given where individuals' own sin makes some contribution to their disease, there are many more where the sins of others or the sinful structures of communities, societies, and political systems threaten countless people's health. At an individual level, for example, we might think of connections between abuse, trauma, and self-destructive behaviors like drug use and eating disorders; at a systemic and structural level, of the massive global health consequences of conflict or economic injustice already alluded to in chapter 3 and under thesis (ix) above.

(xii) *There is a complex relationship between disease, impairment, and disability.*
As is clear from the debates within the disability movement and disability studies literature surveyed in chapter 2, the sharp separation of diseases and disorders from impairments and disabilities, which might be inferred from a simple reading of the social model of disability, cannot be sustained. Many diseases, particularly chronic diseases, give rise to impairments, and as many disability activists and authors have emphasized in critical dialogue with the social model, impairments do sometimes — though not always — cause pain, suffering, and limitation.[70] Numerous authors have likewise emphasized the need to avoid setting up a simplistic opposition between the disability community and the health care professions, and instead to discern the forms of appropriate medical interven-

---

69. A vivid recent literary portrayal of this can be found in the distinctly unattractive character of Professor Michael Beard in Ian McEwan, *Solar* (London: Vintage, 2011).

70. E.g., Liz Crow, "Including All of Our Lives: Renewing the Social Model of Disability," in *Exploring the Divide: Illness and Disability*, ed. Colin Barnes and Geof Mercer (Leeds: Disability Press, 1996), pp. 55-73; Jenny Morris, *Pride Against Prejudice: Transforming Attitudes to Disability* (London: The Women's Press, 1991).

tion and support actually needed by people with particular impairments. The relationship between impairment and disability is also found to be more complex than some versions of the social model have sometimes suggested, leading some disability studies authors such as Shakespeare to move away from the social model itself to a more complex interactional model close to that adopted by the WHO in the *International Classification of Functioning, Disability, and Health (ICF)*.[71]

Having said all that, however, the social model of disability offers crucially important insights for a theological account of health, disease, and human flourishing. One is that what causes a disabled person disadvantage or suffering is often not her impairment, limitation, or loss of function itself, so much as the social, political, or economic conditions that *make* the impairment a problem when it need not be. A second is that social attitudes, assumptions, prejudices, and professional orthodoxies sometimes inappropriately pathologize impairments or generate false assumptions about the suffering those impairments cause.[72] A third is the critical question raised by Amundson and others about the inappropriate use of notions of "normal function" to label unusual ways of fulfilling the goals of embodied creaturely life as abnormal, and therefore pathological.[73]

In light of all these considerations, disability should be thought of as referring to a *larger* domain of human life than the domain of disease, though related to the latter and at least partly inclusive of it. When we are considering what we understand by health and disease, and especially when we have to discern what practical response is called for, at least two critical questions from a disability perspective should be kept in mind. First, when people experience threats or interruptions to their flourishing associated with losses or differences of bodily or mental function, to what extent is their flourishing really threatened by those losses or differences of function, and to what extent by social attitudes and conditions that make life harder for people with such differences? Secondly, are we ever tempted (for example, because of personal or cultural prejudices) to misidentify as *threats* to creaturely flourishing forms of embodied human life that are better understood as unusual *ways* of flourishing as a human creature?

---

71. Tom Shakespeare, *Disability Rights and Wrongs* (London: Routledge, 2006), pp. 54-62; World Health Organization, *International Classification of Functioning, Disability, and Health,* Short Version (Geneva: WHO, 2001).

72. See, e.g., Amundson, "Against Normal Function"; Amundson, "Quality of Life, Disability, and Hedonic Psychology"; Basnett, "Health Care Professionals and Their Attitudes."

73. Amundson, "Against Normal Function."

(xiii) *While a clear distinction between disease and illness cannot be sustained, attempts to make such a distinction draw attention to the divergence between professionals' and patients' perspectives on disease, and attending carefully to the latter brings to the fore some important theological and pastoral questions about the meanings of illness and suffering.*

Several of the philosophical accounts surveyed in chapter 1 draw a clear-cut distinction between *disease,* an objective pathological state or process susceptible to scientific investigation, and *illness,* a patient's subjective experience of a loss of health. K. W. M. Fulford is one author who makes such a distinction, arguing that of the health concepts, illness rather than health or disease has conceptual priority.[74] This distinction draws attention to the relative neglect, in some theories of health, of patients' experiences of illness and suffering. S. Kay Toombs's phenomenological treatment of illness, also surveyed in chapter 1, powerfully addresses that neglect, offering an analysis of the divergence between professionals' and patients' perspectives and a compelling account of the latter.[75]

The theological account proposed in this book, in which health is understood in terms of the capability to realize the proximate ends of embodied, creaturely human life, and diseases are internal states or processes that tend to hinder or threaten the realization of those ends, will probably not support a sharp distinction between disease and illness — certainly not if that distinction depends on a modern kind of fact-value separation. In that respect it will agree with Christopher Megone's Aristotelian account, which defines illness and disease as failures of those functions of the organism that contribute to the realization of a good human life, and denies any sharp distinction between illness and disease.[76] However, while refusing a sharp distinction, this theological account can also accommodate an awareness of the important differences in perspective mapped phenomenologically by Toombs, and certainly it will affirm the importance of attending carefully to patients' perspectives and experiences.

Attention to those perspectives and experiences raises the questions of theodicy and the meaning of suffering briefly discussed in chapter 3. As I noted there, Barth's account of evil as "nothingness" shifts the emphasis

---

74. K. W. M. Fulford, "Praxis Makes Perfect: Illness as a Bridge Between Biological Concepts of Disease and Social Conceptions of Health," *Theoretical Medicine* 14 (1993): 305-320.

75. S. Kay Toombs, *The Meaning of Illness: A Phenomenological Account of the Different Perspectives of Physician and Patient* (Dordrecht: Kluwer Academic, 1992).

76. Christopher Megone, "Mental Illness, Human Function, and Values," *Philosophy, Psychiatry, and Psychology* 7.1 (2000): 45-65 (p. 61).

away from attempts to explain the origins of evil or demonstrate the justice of God, toward a focus on what God has done to overcome evil and what we are called to do in response. In relation to disease, one aspect of the response we are called to make takes the form of the Christian mandate to heal, but another has to do with the sufferer's response to his or her experience of illness.

This aspect is highlighted by the story of Paul's thorn in the flesh (2 Cor. 12:1-10): a profoundly ambiguous and paradoxical narrative of power, weakness, and suffering. As I noted in chapter 3, Paul says that he experienced the "thorn" — whatever it was — as a real and terrible evil, and pleaded with God to be rid of it. Yet having received God's answer, "My grace is sufficient for you, for power is made perfect in weakness," he pronounces himself content even in his sufferings, "for whenever I am weak, then I am strong." Paul's suffering is real evil; yet it has also proved to be an occasion for him to come to know and trust God's goodness more fully. Karl Barth's account of disease has the same double-edged quality. He is unequivocal in naming disease as evil, an aspect of "nothingness," which God has already overcome in the death and resurrection of Christ. Yet without denying this identification at all, he seeks to hold it in tension with the thought that the experience of illness can be a sharp reminder of the finitude of our present life in this world.[77] As such, it can steer us away from the error of investing our ultimate hopes in this mortal life and direct our attention towards the hope of eternal life promised in God's good future.[78]

This is not something that should ever be said glibly. To say that illness

---

77. Barth, *CD* III.4, pp. 372-374.

78. Corinna Delkeskamp-Hayes, from an Orthodox perspective, pursues this line of thought further and more boldly, arguing that disease is one aspect of the corruptibility of human nature that results from Adam's sin, but it is now placed within the horizon of God's salvific activity in the death and resurrection of Jesus Christ. After the resurrection, disease and death persist in the world, partly because God lovingly respects the freedom of human creatures (and therefore does not simply cancel out the consequences of humanity's freely-chosen sin), but also because God turns the evil of disease to our good, making it possible for disease to be a kind of "therapy" for our souls. Physical suffering can not only *remind* sufferers of their sinfulness and mortality, prompting them to repentance and a life of spiritual discipline; "[w]ith the efficacy of Christ's cross secured, such experiences, in their very physically threatening reality, as infringing on the body's passionately cherished integrity, can by themselves produce such a therapeutic effect." Corinna Delkeskamp-Hayes, "Why Patients Should Give Thanks for Their Disease: Traditional Christianity on the Joy of Suffering," *Christian Bioethics* 12.2 (2006): 213-228 (pp. 219-220), citing Jean-Claude Larchet, *The Theology of Illness* (Crestwood, NY: St. Vladimir's Seminary Press, 2002).

can direct us toward God's goodness and be an occasion for the growth of our trust in God is emphatically not to claim simplistically that suffering is good for the soul, or to think of it as a kind of bracing discipline akin to early morning runs and cold showers. To take the latter view would collapse Barth's (and Paul's) dialectic, and fail to do justice to the New Testament witness to the war waged by God in Christ on the destructive forces of chaos and disease. But it is to say that — mysteriously, by God's grace — even such suffering can be transformed. Such transformation of suffering is possible, in Christian perspective, because Christ has shared in our sufferings.[79] Conversely, some Christian reflection points to another kind of transformation of suffering, echoing Paul's statement that "in my flesh I am completing what is lacking in Christ's afflictions for the sake of his body, that is, the church" (Col. 1:24). In his apostolic letter *Salvifici Doloris*, for example, Pope John Paul II wrote of the transformation of human suffering into a sharing in, and completion of, the redemptive suffering of Christ.[80] Others are quite critical of these claims; Nigel Biggar, for example, finds John Paul's exposition obscure in various respects. He does, however, allow that the suffering of disease or illness can *sometimes*, in a weaker sense, play a part in the redemption both of the sufferers and of those around them: our own suffering can prompt us to reorient our lives in the ways already discussed, and such redirected lives can be a prophetic witness to others and the wider community.[81]

79. Delkeskamp-Hayes ("Why Patients Should Give Thanks for Their Disease") claims that the account she presents of the relationship between sin and disease is neglected in Western Christianity, but it is clear from my discussion that there are strong echoes of it in the work of Western theologians such as Barth. That said, there remain differences of emphasis, at least. The account I have developed undoubtedly has space for the recognition that *some* instances of disease and suffering can, by God's grace, be turned to the spiritually "therapeutic" effect for which Delkeskamp-Hayes argues, but I would be more cautious than I think she is about turning this into a generalization. Certainly the *first* thing to be said by Christians about any instance of disease is the indignant naming of it as an evil — an offense against God's goodness — clearly audible in Jesus' response to those who objected to his healing on the Sabbath: "Ought not this woman, a daughter of Abraham whom Satan bound for eighteen long years, be set free from this bondage on the Sabbath day?" (Luke 13:16).

80. John Paul II, *Salvifici Doloris* (1984), available online at http://www.vatican.va/holy_father/john_paul_ii/apost_letters/documents/hf_jp-ii_apl_11021984_salvifici-doloris_en.html.

81. Nigel Biggar, *Aiming to Kill: The Ethics of Suicide and Euthanasia* (London: Darton, Longman, and Todd, 2004), pp. 50-55.

(xiv) *Barth's remark that "health, like life in general, is not an eternal but a temporal and therefore a limited possession"*[82] signals a Christian ambivalence about death and its relationship to disease.

The close association between disease and death makes it easy, in modernity, to identify one with the other simplistically, regarding death simply as a disease that is in principle preventable.[83] This could be seen as the logical conclusion of an influential attitude to medicine articulated early in the modern age by Francis Bacon and others, that to pronounce diseases incurable is to "enact a law of neglect, and exempt ignorance from discredit."[84] The positive influence of this refusal should not be gainsaid: it has provided a powerful impetus for the astonishing achievements of scientific medicine in modern times, which have made life vastly better for those fortunate enough to have access to modern health care. Furthermore, the project to defeat death technologically might appear to claim some support from a theological tradition that regards death as "[t]he last enemy to be destroyed" (1 Cor. 15:26) and looks forward to a future when "[d]eath will be no more; mourning and crying and pain will be no more" (Rev. 21:4).[85]

Yet there are some obvious difficulties with such a view. Biologically, life would not be possible without death, as both evolution and ecology make clear.[86] This observation has prompted some Christians to argue that the Christian association of death with sin must refer to the "spiri-

---

82. Barth, *CD* III.4, p. 371.

83. Aubrey de Grey, for example, makes something close to this equation by denying any distinction between age-related diseases and aging itself, regarding aging as simply the early stages of age-related diseases of which we could eventually die. He believes that there is a strong probability that his SENS (strategies for engineered negligible senescence) approach will eventually make it possible to postpone aging indefinitely: see de Grey and Rae, *Ending Aging*.

84. Francis Bacon, *The Advancement of Learning and The New Atlantis*, The World's Classics (London: Oxford University Press, 1913), p. 123. Available online at http://archive.org/details/advancementnewat00bacouoft.

85. So Michael J. Reiss, "And in the World to Come, Life Everlasting," in Deane-Drummond, ed., *Brave New World?* pp. 49-67.

86. See Fred Simmons, "Contemporary Cosmologies and Christian Faith," n.d., online at http://www.journeyoftheuniverse.org/storage/Fred.Simmons.pdf. Recognizing this, however, need not entail the "tripartite value theory" Simmons proposes as a solution to the problem of theodicy that it raises: see further Christopher Southgate, *The Groaning of Creation: God, Evolution, and the Problem of Evil* (Louisville: Westminster John Knox, 2008), and Neil Messer, "Natural Evil After Darwin," in *Theology After Darwin*, ed. Michael Northcott and R. J. Berry (Carlisle: Paternoster, 2009), pp. 139-154, for two alternatives.

tual" death of alienation from God; there is nothing evil or contrary to God's purposes about biological death in itself.[87] In slightly more nuanced fashion, Frederick Gaiser has argued that there is nothing in the biblical texts to suggest that if humans had not fallen, they would have been immortal. From this he infers that mortality itself is not an evil, and if physical death remained in its proper place at the end of a fulfilled life, it would hold no terror for us. We experience death as an evil in a fallen world only because it will not stay where it is meant to be, but intrudes into the midst of life; only then does it become our enemy.[88]

While each of these opposite perspectives reflects important insights, both are overly simple: neither fully captures the complex dialectic of the Christian tradition's view of death. As Bernd Wannenwetsch argues, that dialectic is better expressed by Dietrich Bonhoeffer's reading of Genesis 3, which offers what Wannenwetsch describes as a "dual coding" of death.[89] In Bonhoeffer's reading of the Fall narrative, the judgment on human sin is that we find ourselves having to live a life in the world that is no longer a gift, but an impossible commandment; this is the "spiritual" death that is not the last, but the *first* enemy defeated by Christ's death and resurrection. Physical death, however, is still an enemy: this is the *last* enemy, to be defeated "when all things are subjected to Christ" (1 Cor. 15:28).

Put differently, Francis of Assisi's phrase "our Sister Bodily Death,"[90] or W. H Draper's well-known paraphrase,

---

87. So, e.g., Arthur Peacocke, "Biology and a Theology of Evolution," *Zygon* 34.4 (1999): 695-712 (p. 700).

88. Frederick J. Gaiser, *Healing in the Bible: Theological Insight for Christian Ministry* (Grand Rapids: Baker Academic, 2010), p. 245. Barth too argues that, although our sin means that in our death we are confronted with the judgment of God, "[t]here is no reason why it should not be an anthropological necessity, a determination of true and natural man, that we shall one day have to die . . .": *CD* III.2, pp. 625-633 (quotation at p. 631). However, his account of the relationship between human finitude and death as judgment is more complex and more akin to the dialectical understanding found in Bonhoeffer, discussed below.

89. Dietrich Bonhoeffer, *Creation and Fall: A Theological Exposition of Genesis 1-3*, Dietrich Bonhoeffer Works, vol. 3, ed. Martin Rüter and Ilse Tödt, trans. John W. De Gruchy and Douglas Stephen Bax (Minneapolis: Fortress, 1997), pp. 131-136; Bernd Wannenwetsch, "From *Ars Moriendi* to Assisted Suicide: Bonhoefferian Explorations into Cultures of Death and Dying," *Studies in Christian Ethics* 24.4 (2011): 428-440.

90. Francis of Assisi, *The Canticle of the Sun*, trans. Bill Barrett (n.p., n.d.). Available online at http://ww2.webster.edu/~barrettb/canticle.htm.

> And thou, most kind and gentle death,
> waiting to hush our latest breath,
> O praise him, alleluia![91]

taken out of context and repeated glibly, give a one-sided impression of the Christian tradition's attitude toward death. It is only because Jesus Christ has been through death and risen again that it can become "Sister Death"; only because "Jesus lives!" that

> [h]enceforth is death
> but the entrance-gate of life immortal.[92]

This is the kind of *defeat* that physical death receives at the hands of Christ: to be transformed from our last enemy to our "Sister," the "entrance-gate to life immortal."

In short, the defeat of our last enemy, death, does not mean an heroic technological effort to cure us of it, as de Grey imagines; that would be better understood as a counterfeit or parody of its true defeat, accomplished by Christ.[93] Nor, on the other hand, does it entail embracing death over-eagerly as a friend, or seeking to meet it on our own terms and bring it under our control, attitudes diagnosed in contemporary Western societies by commentators such as Wannenwetsch. In Christian terms, the defeat of death is accomplished neither by seeking it nor by evading it for as long as possible, but by being ready to meet it, when it comes to us, "in sure and certain hope of the Resurrection to eternal life, through our Lord Jesus Christ," to use Cranmer's resonant phrase.[94] As Stanley Hauerwas remarks (with perhaps a touch of hyperbole), "Christians are not fundamentally concerned about living. Rather, their concern is to die for the right thing."[95] I observed in chapter 3 that there is a "kind of death by which [we may] glorify God" (John 21:19); but this is not the same as the

---

91. W. H. Draper, "All Creatures of Our God and King," in *Rejoice and Sing* (Oxford: Oxford University Press, 1991), no. 79, v. 6.

92. C. F. Gellert, "Jesus Lives! Thy Terrors Now," in *Rejoice and Sing*, no. 239, v. 4.

93. *Pace* Reiss, "And in the World to Come, Life Everlasting."

94. "The Order for the Burial of the Dead," in *Book of Common Prayer* (1662), available online at http://www.churchofengland.org/prayer-worship/worship/book-of-common-prayer/at-the-burial-of-the-dead.aspx.

95. Stanley Hauerwas, *Suffering Presence: Theological Reflections on Medicine, the Mentally Handicapped, and the Church* (Edinburgh: T&T Clark, 1988), p. 92.

dominant ideals of a good death on offer in late modern Western societies,[96] nor as the kinds of heroic death that might have been admired in classical antiquity: the kind of death that participates in the defeat of death is that of a saint, not a hero.[97]

Practical Implications

(xv) *The divine command to "will to be healthy" (Barth) signals a clear, albeit qualified, theological affirmation of the work of medicine and healing. Since genuine health and healing are always God's gift, a false opposition between medicine and the Christian healing ministry should be resisted, though the latter properly has a wider field of concern than the former.*

I have already noted the Christian mandate to heal. This follows from God's command to "will to be healthy" or to live the kind of creaturely life that God has given us to live, and from its corollary, the command to resist diseases and other threats to creaturely flourishing, following the example of the Gospel healing narratives which show how "[God] Himself has already marched against that realm on the left, and that He has overcome and bound its forces and therefore those of sickness in Jesus Christ and His sacrifice, by which the destroyer was himself brought to destruction."[98] This leads Barth to echo Jesus ben Sirach's call to "honor the physician": the Christian tradition supports a very positive assessment of the vocation of medicine and the work of healing.[99]

This Christian affirmation of medicine, however, comes with some caveats and qualifications. First, as Barth remarks, it has an important *but limited* role in supporting health. He warns against the conceit that "the doctor is the one who really heals," because health, like life itself, is God's gift: the role of medicine is to remove or circumvent obstacles to health and to palliate the effects of those that cannot be overcome or circumvented.[100] Secondly, alongside ben Sirach's call to "honor the physician" are more critical strands of biblical tradition such as the story of King Asa,

---

96. Cf. Joel Shuman and Brian Volck, *Reclaiming the Body: Christians and the Faithful Use of Modern Medicine* (Grand Rapids: Brazos, 2006), pp. 7-8.

97. Samuel Wells, "The Disarming Virtue of Stanley Hauerwas," *Scottish Journal of Theology* 52.1 (1999): 82-88.

98. Barth, *CD* III.4, p. 368.

99. Sirach 38:1-15; Barth, *CD* III.4, pp. 360-363.

100. Barth, *CD* III.4, p. 361.

castigated by the Chronicler for relying on physicians rather than the Lord (2 Chron. 16:12), and Mark's observation that the woman with the flow of blood "had endured much under many physicians, and had spent all that she had; and she was no better, but rather grew worse" (5:26). While strong affirmations of the profession of medicine can be found in the tradition, they are not the only voices. The warier ones at any rate emphasize the limits to the power of medicine and to what can and should be expected of it; they warn us against setting up doctors as alternative saviors. Thirdly, as commentators such as Gerald McKenny argue, modern medicine is distinctive not only in the scope of its understanding of disease and ability to treat it, but also in its moral character and commitments.[101] The "Baconian project," as McKenny calls it, represents the aspiration to set humanity free from finitude and chance, harnessing technology for the relief of suffering and the expansion of individual choice. While — as he is careful to emphasize — there is no reason to deny the great benefits won by technological medicine or to hanker after some pre-technological golden age, Christians do have good reason to be critical of any "Baconian" aspirations or commitments that they encounter in aspects of contemporary medical practice, as suggested in chapter 3.

In that chapter, I also noted the uneasy relationship between the Christian ministry of healing and the professional world of healthcare. However, I suggested that aspects of this unease were encouraged by false oppositions between divine and human agency and between scientific medicine and miracle. Properly understood, there is no opposition between healing by means of medicine and the Christian ministry of healing, as most authors on the healing ministry are quick to say — even if they do not always follow this through consistently. As Barth and others insist, all true health and true healing are God's gift, and healing is no less God's work if it is accomplished by means of the human agency of health professionals. This argues for cooperation between health professionals and practitioners of the Christian ministry of healing with as little mutual suspicion as possible, but also without dividing up the territory of healing between medicine and religion. Prayer for healing is not an alternative to surgery, chemotherapy, or antibiotics, but neither is it irrelevant to those conditions for which surgery, chemotherapy, or antibiotics are indicated.

That said, the Christian ministry of healing properly has a wider field

---

101. Gerald P. McKenny, *To Relieve the Human Condition: Bioethics, Technology, and the Body* (Albany: State University of New York Press, 1997).

of concern than the work of health care professionals. It is an aspect of Christian pastoral care, and pastoral care is properly concerned for all aspects of a person's life: her life before God, her relationships with others, her strength to live a human life and the living out of her vocation in a particular place and time. The boundaries between these aspects of human flourishing (or the Creator's liberating command) are hardly clear-cut, so pastoral care that includes the ministry of healing can properly speak of "wholeness" and be concerned with the "whole person," even though it is unwise and potentially misleading to equate "health" with "wholeness" or "healing" with "making whole."

(xvi) *Not everyone will be cured, and even those who are will die sooner or later — not only because medical skill and knowledge are incomplete, but because embodied life in this world is finite. The task of healing (including medicine) therefore includes the call to continue caring when cure is no longer possible. The life of Christian communities should offer resources that can support this work of care and transform the way suffering is understood and experienced.*

Barth, as we have seen, emphasizes that "health, like life in general, is not an eternal but a temporal and therefore a limited possession,"[102] and holds that diseases or illnesses can direct our attention beyond the horizon of this mortal life towards the ultimate hope promised by God in Christ. To recognize that not all diseases will ever be cured, and that despite all that medicine can do we remain mortal, need not therefore be a capitulation in the struggle against disease. It can reflect a proper appreciation of the relationship between ultimate and penultimate human ends. Because of this, the ministry of healing and the calling of health care professionals must — of course — be concerned with resisting diseases. However, as Hauerwas argues, they must also be concerned with continuing to care when cures are no longer possible.[103] The calling to be present to those who cannot be cured can be powerfully informed by testimonies such as Toombs's phenomenological account of illness and the humane and insightful narratives of practitioners such as Oliver Sacks;[104] in chapter 3 I

---

102. Barth, *CD* III.4, p. 371.
103. Hauerwas, *Suffering Presence*, pp. 63-83.
104. Toombs, *The Meaning of Illness;* Oliver Sacks, *The Man Who Mistook His Wife for a Hat and Other Clinical Tales* (New York: Summit, 1985); Oliver Sacks, *The Mind's Eye* (London: Picador, 2011).

also noted Hauerwas's claim that to sustain such practices of care, "something very much like a church is needed."[105]

Likewise, part of what makes assisted dying appear such an attractive and important option in many late modern cultures seems to be the fear that our deaths will be accompanied by pain, suffering, indignity, and the loss of autonomy, control, and even self. In the face of such suffering and loss it is tempting to try and take as much control as possible over the time and manner of our own dying. If a Christian narrative of resurrection and the hope of eternal life is to challenge such a response to these fears, then it cannot be merely words. The onus is on Christian communities, which year by year live through the story of Good Friday and Easter, to live the kind of common life through which suffering and dying people can find the hope, courage, and other resources they need to endure such pain, indignity, and loss. Participation in the life of the Church should train us, as Hauerwas and Willimon argue, to *see* differently:[106] to develop a vision that transforms the way in which we understand, perceive, *and experience* suffering and death.

---

105. Hauerwas, *Suffering Presence*, p. 65.
106. Hauerwas and Willimon, *Resident Aliens*, p. 95.

Conclusion

The theological account of health, disease, and illness that I have attempted to develop in this book is summarized in the following sixteen theses, which were set out, explained, and defended in chapter 4.

Humans as Creatures

(i) A theological account of health and disease must begin by recognizing that we are creatures of a particular, human, kind; what this means, we understand theologically in the light of Jesus Christ. As creatures of a particular kind, human beings have both proximate and ultimate ends, which are God's gift to us in creating us as beings of this kind. Our good or flourishing consists in the fulfillment of these ends.

(ii) Our ultimate end is life in union with God, which means that theological talk of human goods and ends (including health) must be placed within an eschatological horizon; but some time-honored Christian ways of understanding that horizon require critical theological reappraisal.

(iii) There are penultimate or proximate ends that are both objective and universal, given simply by virtue of our being creatures of this particular, human, kind. Knowledge of these ends comes in the first place from God's revelation in Christ, rather than any empirical study of the human species, but many kinds of human knowledge and insight can be critically assimilated to this theological understanding.

(iv) As well as those penultimate goods, goals, and ends common to all human creatures, there are goals and ends that are particular and contextual, aspects of the vocations of particular human lives. Discernment of both these particular vocational goals and the appropriate routes towards universal ends in particular contexts will require habits of attentiveness to God's command, or, put another way, the virtue of practical wisdom or prudence.

Health and Creaturely Flourishing

(v) Health is not to be equated with the whole of human well-being (as in the World Health Organization definition), and Christian talk of "wholeness" should be treated with considerable caution. Nonetheless, there are important insights expressed in talk of wholeness that can and should be articulated in other ways.

(vi) Karl Barth's definition of health as "the strength for human life"[1] expresses some themes that are at the very heart of a theological understanding of health: it recognizes the character of embodied human life as a divine gift that comes with a divine call, it offers an account of physical and mental integration that captures another important insight expressed in Christian talk of wholeness, and it allows a balanced understanding of health as the "capability, vigor, and freedom" to live a human life.

(vii) A Christocentric account of creaturely flourishing can be hospitable to evolutionary and other biological insights about natural functions, though these insights, like those from other sources, must be *critically* appropriated.

(viii) Health is a *penultimate* good, concerned with the fulfillment of proximate human goals or ends, although it points toward the ultimate fulfillment of human life. If it is treated as an ultimate end — which is a form of idolatry — the results are likely to prove destructive.

(ix) Health and the "will to be healthy" are not merely individual, but also social and political matters.

Disease, Suffering, Evil, and Sin

(x) Diseases are to be understood as a particular class of threats to creaturely flourishing: internal states or processes that tend to disrupt or

---

1. Karl Barth, *Church Dogmatics*, 13 vols., ed. and trans. Geoffrey W. Bromiley and Thomas F. Torrance (Edinburgh: T&T Clark, 1956-75), III.4, p. 356. Hereafter *CD*.

threaten the fulfillment of the proximate ends of embodied creaturely human life. However, discrimination between diseases and non-diseases cannot be entirely a matter of clear definition, and will require the exercise of practical wisdom.

(xi) There is a complex but important set of connections to be made between disease, evil, and sin.

(xii) There is a complex relationship between disease, impairment, and disability.

(xiii) While a clear distinction between disease and illness cannot be sustained, attempts to make such a distinction draw attention to the divergence between professionals' and patients' perspectives on disease, and attending carefully to the latter brings to the fore some important theological and pastoral questions about the meanings of illness and suffering.

(xiv) Barth's remark that "health, like life in general, is not an eternal but a temporal and therefore a limited possession"[2] signals a Christian ambivalence about death and its relationship to disease.

Practical Implications

(xv) The divine command to "will to be healthy" (Barth) signals a clear, albeit qualified, theological affirmation of the work of medicine and healing. Since genuine health and healing are always God's gift, a false opposition between medicine and the Christian healing ministry should be resisted, though the latter properly has a wider field of concern than the former.

(xvi) Not everyone will be cured, and even those who are will die sooner or later — not only because medical skill and knowledge are incomplete, but because embodied life in this world is finite. The task of healing (including medicine) therefore includes the call to continue caring when cure is no longer possible. The life of Christian communities should offer resources that can support this work of care and transform the way suffering is understood and experienced.

If the theological tradition in which I have been working ought to understand health and disease in something like this way, what are the implications of that understanding for the practical ethical issues and examples with which the book began?[3]

2. Barth, *CD* III.4, p. 371.
3. Parts of the following discussion are taken, with modifications, from Neil Messer,

The Therapy/Enhancement Distinction

The first thing to say is that this theological account of health and disease enables us to make a distinction between therapy and enhancement that can do real conceptual and ethical work. Health makes possible the fulfillment of certain penultimate goods, goals, and ends that have to do with living an embodied, creaturely life of a human kind. Diseases are obstacles to the fulfillment of those goals, and therapeutic interventions are those that remove, or circumvent, or mitigate the effects of, such obstacles.[4] For example, on this account an intervention to reverse the effects of the genetic mutation that causes Duchenne muscular dystrophy would be recognizably therapeutic in a way that "gene doping" to build up muscle strength and enhance athletic performance would not.

However, while my account suggests that there is a real and morally significant distinction between therapy and enhancement, the line between the two is by no means sharp or clear. Some examples are clearly on one or other side of the line: genetic interventions to replace the mutant gene causing cystic fibrosis or to remove a genetic predisposition to some form of cancer would be clear examples of therapy, whereas the attempts at "radical life extension" advocated by Aubrey de Grey and others fall equally clearly into the category of enhancement: in the theological perspective articulated here, aging is not a disease. But other cases are more ambiguous. It is highly questionable whether various conditions that have conventionally been regarded in modern times as medical problems, such as inherited forms of intellectual disability or deafness, should be so described. If not, then genetic interventions directed at those conditions (should such interventions ever be technically possible) could not, of course, be called "therapies." This is a clear case where practical wisdom, or the habits and skills of attentiveness to God's command, will be needed to discern which interventions are properly considered therapeutic. The account of health developed in this book might suggest criteria that would go some way to resolving these ambiguities. A key question to ask of any proposed genetic intervention would be: Is the condition that it is intended to address a genuine threat to the "strength for human life," theologically understood? But no theological criterion, in isolation from the

---

"Toward a Theological Understanding of Health and Disease," *Journal of the Society of Christian Ethics* 31.1 (2011): 161-178 (pp. 172-174).

4. Barth, *CD* III.4, pp. 361-363.

life and practice of the Christian community, will entirely resolve the ambiguity. I have already claimed, with reference to Hauerwas and Willimon, that the discernment of genuine threats to the "strength for life" and the appropriate response to those threats requires a Christian character formed by the Church's distinctive narratives and practices. Their story of Dorothy suggests that her church was able to discern what others did not, that she was not "ill" or "afflicted" just by virtue of having Down syndrome, because they read texts like Matthew 18:1-4 and met around the Lord's table. A transformed vision formed by Word and sacraments is indispensable for the kind of discernment that is needed here, though such a transformed vision can certainly be informed by theological reflection that suggests the right questions to ask.

Indeed, critical theological reflection on disability might press the point further and suggest that "therapy" and "enhancement" are not the only possible categories of clinical intervention to consider. There could also be forms of intervention — perhaps well-intentioned and *understood* as therapies by those who practice them — which would be better understood as suppressing diversity. The giving of cochlear implants to profoundly deaf young children would be understood in this way by many Deaf activists.

Moreover, even when genetic conditions are recognized as genuine threats to the "strength for human life," there are further questions to be asked about the kinds of intervention, genetic or otherwise, that can be counted as morally legitimate ways of meeting those threats. An intervention that depended in some way on the use of human embryonic stem cells, for example, would raise serious questions regardless of whether it counted as a therapy.[5]

Finally, the eschatological horizon of this account of health and disease offers a critical perspective on some projects of human enhancement — certainly those that have the aim of perfecting human nature or transcending all human limits, such as the kind of "transhumanist" project whose goal is effectively to defeat death.[6] The kinds of this-worldly human flourishing that we call "health" are real, important, and worth cherishing and

---

5. For discussion of these questions, see Neil Messer, *Respecting Life: Theology and Bioethics* (London: SCM, 2011), ch. 4, and Brent Waters and Ronald Cole-Turner, eds., *God and the Embryo: Religious Voices on Stem Cells and Cloning* (Washington, DC: Georgetown University Press, 2003).

6. See further Brent Waters, *From Human to Posthuman: Christian Theology and Technology in a Postmodern World* (Aldershot: Ashgate, 2006).

promoting as penultimate goods, but they betray us if we try to pin ultimate hopes onto them, or turn them into a secular form of salvation.

Norman Daniels's Resource Allocation Theory

The second of my opening issues was the allocation of health care resources, and in particular the role that concepts of health and disease play in Norman Daniels's contractarian approach.[7] As we saw, Daniels needs an objective, non-normative account of health and disease to underpin his argument that health needs have special moral importance because of their role in safeguarding citizens' fair shares of the normal opportunity range. In the first presentation of his approach, he relied on Christopher Boorse's biostatistical model to supply this objective account.[8] More recently — partly in response to criticisms of Boorse — he has favored Wakefield's model of disorders as "harmful dysfunctions," arguing that the "dysfunction" component can support a non-normative account of normal functioning.[9]

However, there are reasons to doubt whether it is possible to give the non-normative account of normal functioning that Daniels requires. Critics such as Anita Silvers and Ron Amundson have argued that purportedly objective, value-free accounts in fact conceal value-laden judgments about what counts as "normal" functioning.[10] I suggested in chapter 2 that Wakefield's model of disorder can be articulated in a way that is compatible with Amundson's account, but only because the conceptual heavy lifting in Wakefield's model is not done by notions of normality. Since it is precisely "normality" that does the work in Daniels's theory, however, his contractarian approach remains vulnerable to Amundson's and Silvers's critiques.

Furthermore, a non-normative account of natural function seems to

---

7. Norman Daniels, *Just Health: Meeting Health Needs Fairly* (Cambridge: Cambridge University Press, 2008), ch. 2.

8. Norman Daniels, *Just Health Care* (Cambridge: Cambridge University Press, 1985), pp. 19-35.

9. Daniels, *Just Health*, pp. 36-42.

10. Anita Silvers, "A Fatal Attraction to Normalizing: Treating Disabilities as Deviations from 'Species-Typical' Functioning," in *Enhancing Human Traits: Ethical and Social Implications*, ed. Erik Parens (Washington, DC: Georgetown University Press, 1998), pp. 95-123; Ron Amundson, "Against Normal Function," *Studies in History and Philosophy of Biological and Biomedical Sciences* 31.1 (2000): 33-53.

depend on a modern kind of fact-value dichotomy. Such a dichotomy was already called into question in chapter 1 by Christopher Megone's Aristotelian account of health, disease, and illness. In later chapters this critique was intensified in the development of a theological teleology, in which a human creature is a being whose good consists in the fulfillment of certain characteristic ends. This perspective yields an understanding of health that is purportedly objective but also avowedly normative. What counts as human health is given in the structure of our kind of creaturely life, which is recognizable theologically as *good* because it reflects the good purposes of the Creator whose gift it is.

In the theological perspective I have developed, Daniels's theoretical reasons for affirming the special moral importance of health are called into question; but that does not require us to *deny* the importance of health. I have argued that it is a real, albeit penultimate, good: in Karl Barth's terms, the "strength for life," or the God-given ability to answer the Creator's call and live a flourishing human life. This gives a strong theological mandate for the practice of health care, and for other activities aimed at protecting and promoting health and resisting those things that threaten it. Since all human beings are addressed by God's liberating command to "will to be healthy,"[11] there are theological reasons to agree with Daniels about the moral importance of a certain level of health care provision. Elsewhere I have argued, drawing on Barth and Bonhoeffer, that in theological perspective the state is called to maintain the conditions in which its citizens are able to receive and respond to God's liberating command.[12] This account has a good deal in common with the concept of the *common good* in Catholic social teaching: the sum total of conditions that enable all to flourish as fully as possible. Such a view of the state's responsibility for the common good might well endorse Daniels's call for "universal coverage . . . for an array of 'decent' or 'adequate' services."[13]

Moreover, one of the other features of Daniels's account in *Just Health* is his recognition that justice requires attention not only to health care provision, but more broadly to the social determinants of health.[14] One need not adopt either his view of normal functioning or his Rawlsian account of justice as fairness to agree wholeheartedly with this practical

---

11. Barth, *CD* III.4, p. 356.
12. Messer, *Respecting Life*, pp. 50-54.
13. Daniels, *Just Health*, p. 143.
14. Daniels, *Just Health*, ch. 3.

claim. My thesis (ix) states that "health and the 'will to be healthy' are not merely individual, but also social and political matters," and I quoted earlier Barth's remark that "Where some are necessarily ill, the others cannot with good conscience will to be well."[15] In short, many of the practical conclusions for which Daniels argues in the field of health care resource allocation might be supported by my account — albeit for different and distinctively theological reasons.

Quality of Life

My third opening issue was concerned with the concept of quality of life, and the uses to which it is put in a wide range of bioethical arguments, including abortion, neonatal care, resource allocation, end-of-life care, and assisted dying.

In relation to the QALY approach to resource allocation and the use of health-related quality of life (HRQoL) measures, it would be tempting to write off the whole project from the outset, on the grounds that the form taken by the "will to be healthy" is a matter of the command of God in each particular situation and cannot be adequately expressed in a standardized scale of measurement drawn up in advance. However, this dismissal would be too sweeping a judgment. The theological perspective I have outlined certainly calls for a proper sense of the provisionality of our judgments about health, but it does not dismiss the possibility of making such judgments with all due caution. It does, however, suggest various caveats and critical questions. First, it warns against over-confidence that any standardized measure — at best an abstraction — can capture the full reality of health. It might also therefore lead us to be more skeptical about some measures and some applications of QoL than others. The use of disease-specific scales of measurement to assess the effects of treatments in individual patients or compare the effectiveness of alternative treatment regimes within particular patient groups, for example, seems less problematic than the use of wide-ranging generic measures to compare treatment effectiveness and cost-effectiveness across widely differing areas of practice.

Second, my account will prompt critical questions about the particular criteria used in QoL scales. It will encourage us to be alert to critiques from disability studies about the assumptions built into QoL scales and the bias

---

15. Barth, *CD* III.4, p. 363.

in some interpretations of QoL data. Do the scales themselves reflect biased assumptions about disabled people's QoL, for example by rating walking (however limited) more highly than wheelchair use (however agile)? Does the disparity between first-person reports and third-person estimates of disabled people's QoL reflect "disablist" prejudices in professional communities or society at large?[16] But more than this, my account encourages *theologically* critical questions about the criteria used to assess QoL. Do all of the criteria used in standard indices really reflect aspects of the "strength for human life" and threats to that strength, or do some of them reflect theologically questionable assumptions about human functioning and flourishing?

Third, while my account can with due caution countenance the use of QoL measures to assess such things as a patient's well-being or the effectiveness of a treatment regime, it will raise strong objections to the use of QoL criteria to assess the *worth* of a human life, whether one's own or someone else's. In this theological perspective, every human life is God's good gift in creation, the object of God's saving love revealed in the incarnation, passion, and resurrection of Christ, and heir to God's promised good future; this is what underwrites the worth of each one. With God's gift and promise comes God's liberating call to respect and protect our own and others' lives. In the sphere of health, that call takes the form of the command to "will to be healthy," as Barth puts it:[17] to receive and foster that "strength for life" which is itself God's gracious gift. If we understand the value of life in this way, we cannot judge the worth of our own or our neighbors' lives by the level of their abilities, capacities, or potentialities. More controversially (in many contemporary Western contexts) this perspective also holds us back from judging that our own or others' lives are so filled with pain or suffering that they are no longer worth living. In so doing, as I observed in thesis (xvi), it poses the strongest of challenges to the Church to be the kind of community that can offer suffering people the resources to endure what would otherwise be unendurable.

---

16. For a complaint of bias in QoL scales, see David Pfeiffer, "The ICIDH and the Need for Its Revision," *Disability and Society* 13.4 (1998): 503-523 (p. 512); on the disparity between first- and third-person estimates of disabled people's QoL, see Ron Amundson, "Quality of Life, Disability, and Hedonic Psychology," *Journal for the Theory of Social Behavior* 40.4 (2010): 374-392, and Ian Basnett, "Health Care Professionals and Their Attitudes Toward and Decisions Affecting Disabled People," in *Handbook of Disability Studies*, ed. Gary L. Albrecht, Katherine D. Seelman, and Michael Bury (Thousand Oaks, CA: Sage, 2001), pp. 450-467.

17. Barth, *CD* III.4, p. 356.

Conclusion: The Value of Theological Plumbing

Although questions about the meaning of health, disease, and illness — particularly the philosophical arguments discussed in chapter 1 — can at times seem arcane and abstruse, it is clear from the foregoing discussion that our answers to these questions have all kinds of concrete implications for important areas of human life, notably in the practice of health care. These discussions are an exercise in what, paraphrasing Mary Midgley, we might call "conceptual plumbing" — intellectual work whose results are usually hidden from view, but nonetheless have important consequences for life above the floorboards, as it were.[18] When we ask *theologically* what we should understand by health, disease, and illness, it quickly becomes apparent that our answers depend on some of the deepest of Christian convictions about human life before God and in the world God has made. There is, of course, much more yet to be said about all this. However, I hope that this book has at any rate demonstrated that a sound theological understanding of what it means to *flourish* as human creatures of God is an indispensable guide to Christian life and action in God's world — including the development of Christian perspectives on health care and on the difficult and contested questions of bioethics.

---

18. Cf. Mary Midgley, *Utopias, Dolphins, and Computers: Problems in Philosophical Plumbing* (London: Routledge, 1996), pp. 1-14.

Bibliography

Abberley, Paul. "The Concept of Oppression and the Development of a Social Theory of Disability." *Disability, Handicap, and Society* 2.1 (1987): 5-19.

Admirand, Peter. "Traversing Towards the Other (Mark 7:24-30): The Syrophoenician Woman Amidst Voicelessness and Loss." In *The Bible: Culture, Community, and Society*. Edited by Angus Paddison and Neil Messer. London: T&T Clark, 2013.

Albrecht, Gary L., and Patrick J. Devlieger. "The Disability Paradox: High Quality of Life against All Odds." *Social Science and Medicine* 48.8 (1999): 977-988.

Altman, Barbara M. "Disability Definitions, Models, Classification Schemes, and Applications." In *Handbook of Disability Studies*. Edited by Gary L. Albrecht, Katherine D. Seelman, and Michael Bury. Thousand Oaks, CA: Sage, 2001.

American Psychiatric Association. *Diagnostic and Statistical Manual of Mental Disorders, Fifth Edition (DSM-5)*. Arlington, VA: American Psychiatric Publishing, 2013.

———. *DSM-5 Development*. http://www.dsm5.org/Pages/Default.aspx.

Amundson, Ron. "Against Normal Function." *Studies in History and Philosophy of Biological and Biomedical Sciences* 31.1 (2000): 33-53.

———. "Quality of Life, Disability, and Hedonic Psychology." *Journal for the Theory of Social Behavior* 40.4 (2010): 374-392.

Ananth, Mahesh. *In Defense of an Evolutionary Concept of Health: Nature, Norms, and Human Biology*. Aldershot: Ashgate, 2008.

Aristotle. *Categories*. Translated by E. M. Edgehill. http://classics.mit.edu/Aristotle/categories.html.

———. *Eudemian Ethics*. http://www.perseus.tufts.edu/hopper/text?doc=Perseus%3Atext%3A1999.01.0050%3Abook%3D1%3Asection%3D1214a.

———. *Metaphysics*. Translated by W. D. Ross. http://www.classicallibrary.org/aristotle/metaphysics/index.htm.

———. *Nicomachean Ethics*. Translated by W. D. Ross. http://classics.mit.edu/Aristotle/nicomachean.1.i.html.

———. *Physics*. Translated by R. P. Hardie and R. K. Gaye. http://classics.mit.edu/Aristotle/physics.2.ii.html.

Arnhart, Larry. "The Darwinian Moral Sense and Biblical Religion." In *Evolution and Ethics: Human Morality in Biological and Religious Perspective*. Edited by Philip Clayton and Jeffrey Schloss. Grand Rapids: Eerdmans, 2004.

Arras, John D., and Elizabeth M. Fenton. "Bioethics and Human Rights: Access to Health-Related Goods." *Hastings Center Report* 39.5 (2009): 27-38.

Augustine, *The Enchiridion*. Translated by J. F. Shaw. *Nicene and Post-Nicene Fathers*, series 1, vol. 3. Edited by Philip Schaff. Grand Rapids: Eerdmans, 1956.

Bacon, Francis. *The Advancement of Learning and the New Atlantis*. The World's Classics. London: Oxford University Press, 1913.

Barnes, Colin. "Disability Studies: New or Not So New Directions?" *Disability and Society* 14.4 (1999): 577-580.

Barrett, C. K. *The Second Epistle to the Corinthians*. London: A&C Black, 1973.

Barth, Karl. *Church Dogmatics*. 13 vols. English translation edited by Geoffrey W. Bromiley and Thomas F. Torrance. Edinburgh: T&T Clark, 1956-75.

Basnett, Ian. "Health Care Professionals and Their Attitudes toward and Decisions Affecting Disabled People." In *Handbook of Disability Studies*. Edited by Gary L. Albrecht, Katherine D. Seelman, and Michael Bury. Thousand Oaks, CA: Sage, 2001.

Bayer, Ronald. *Homosexuality and American Psychiatry: The Politics of Diagnosis*. New York: Basic Books, 1981.

Beasley-Murray, George R. *John*. Word Biblical Commentary 36. Waco: Word, 1987.

Begum, Nasa, and Gerry Zarb. "Measuring Disabled People's Involvement in Local Planning." Measuring Disablement in Society Working Paper 5. Leeds: Disability Archive, 1996.

Biggar, Nigel. *Aiming to Kill: The Ethics of Suicide and Euthanasia*. London: Darton, Longman, and Todd, 2004.

———. *Behaving in Public: How to Do Christian Ethics*. Grand Rapids: Eerdmans, 2011.

———. *The Hastening That Waits: Karl Barth's Ethics*. Oxford: Clarendon, 1993.

———. "Karl Barth's Ethics Revisited." In *Commanding Grace: Studies in Karl Barth's Ethics*. Edited by Daniel L. Migliore. Grand Rapids: Eerdmans, 2010.

Bonhoeffer, Dietrich. *Creation and Fall: A Theological Exposition of Genesis 1–3*. Dietrich Bonhoeffer Works, vol. 3. Edited by Martin Rüter, Ilse Tödt, and John W. De Gruchy, translated by Douglas Stephen Bax. Minneapolis: Fortress, 1997.

———. *Ethics*. Dietrich Bonhoeffer Works, vol. 6. Edited by Ilse Tödt, Heinz Eduard Tödt, Ernst Feil, and Clifford J. Green, translated by Reinhard Krauss, Charles C. West, and Douglas W. Stott. Minneapolis: Fortress, 2005.

———. *Letters and Papers from Prison*. Edited by Eberhard Bethge, translated by Reginald Fuller, Frank Clark, John Bowden, et al. London: SCM, 1971.

———. "Stationen der Freiheit." 1944. http://www.luther2017.de/465-dietrich-bonhoeffer-stationen-der-freiheit.

Boorse, Christopher. "Disability and Medical Theory." *Philosophical Reflections on Disability*. Philosophy and Medicine 104. Edited by D. Christopher Ralston and Justin Ho. New York: Springer, 2010.

———. "Health as a Theoretical Concept." *Philosophy of Science* 44.4 (1977): 542-573.

———. "On the Distinction Between Disease and Illness." *Philosophy and Public Affairs* 5.1 (1975): 49-68.

———. "A Rebuttal on Health." In *What Is Disease?* Edited by James M. Humber and Robert F. Almeder. Totowa, NJ: Humana Press, 1997.

———. "What a Theory of Mental Health Should Be." *Journal for the Theory of Social Behavior* 6 (1976): 61-84.

Boyle, Joseph. "Limiting Access to Health Care: A Traditional Roman Catholic Analysis." In *Allocating Scarce Medical Resources: Roman Catholic Perspectives*. Edited by H. Tristram Engelhardt Jr. and Mark J. Cherry. Washington, DC: Georgetown University Press, 2002.

Braddock, David L., and Susan L. Parish. "An Institutional History of Disability." In *Handbook of Disability Studies*. Edited by Gary L. Albrecht, Katherine D. Seelman, and Michael Bury. Thousand Oaks, CA: Sage, 2001.

Brock, Brian. Review of Celia Deane-Drummond, ed., *Brave New World: Theology, Ethics and the Human Genome Project* and Celia Deane-Drummond, Bronislaw Szerszynski, and Robin Grove-White, eds., *Re-Ordering Nature: Theology, Society and the New Genetics*. *Studies in Christian Ethics* 19.1 (2006): 110-116.

———. "Why the Estates? Hans Ulrich's Recovery of an Unpopular Notion." *Studies in Christian Ethics* 20.2 (2007): 179-202.

Brock, Brian, and Stephanie Brock. "Being Disabled in the New World of Genetic Testing: A Snapshot of Shifting Landscapes." In *Theology, Disability, and the New Genetics: Why Science Needs the Church*. Edited by John Swinton and Brian Brock. London: T&T Clark, 2007.

Buller, Cornelius A. "Healing Hope: Physical Healing and Resurrection Hope in a Postmodern Context." *Journal of Pentecostal Theology* 10.2 (2002): 74-92.

Bury, Michael. "A Comment on the ICIDH2." *Disability and Society* 15.7 (2000): 1073-1077.

———. "Defining and Researching Disability." In *Exploring the Divide: Illness and Disability*. Edited by Colin Barnes and Geof Mercer. Leeds: Disability Press, 1996.

———. *Health and Illness in a Changing Society*. London: Routledge, 1997.

Callahan, Daniel. "The WHO Definition of 'Health.'" *Hastings Center Studies* 1.3 (1973): 77-88.

Caspary, Almut. *In Good Health: Philosophical-Theological Analysis of the Concept of Health in Contemporary Medical Ethics*. Stuttgart: Franz Steiner, 2010.

Cassell, Eric J. *The Nature of Suffering and the Goals of Medicine*. 2nd ed. New York: Oxford University Press, 2004.

Chappell, Anne Louise. "From Normalization to Where?" In *Disability Studies: Past, Present, and Future*. Edited by Len Barton and Mike Oliver. Leeds: Disability Press, 1997.

Charlton, James I. *Nothing about Us without Us: Disability Oppression and Empowerment*. Berkeley: University of California Press, 2000.

Chiquete, Daniel. "Healing, Salvation, and Mission: The Ministry of Healing in Latin American Pentecostalism." *International Review of Mission* 93.370/371 (2004): 474-485.

Church of England. "The Order for Morning Prayer Daily Throughout the Year." In *The Book of Common Prayer* (1662).

———. "The Order for the Burial of the Dead." In *The Book of Common Prayer* (1662).

Clements, Keith. Transcript of untitled broadcast sermon. BBC Radio 4, 25 March 2012. http://www.bbc.co.uk/programmes/b01dtfsp.

Cochrane, James R. "Religion, Public Health, and a Church for the 21st Century." *International Review of Mission* 95.376/377 (2006): 59-72.

Corker, Mairian. "Deafness/Disability: Problematizing Notions of Identity, Culture, and Structure." In *Disability, Culture, and Identity*. Edited by Sheila Riddell and Nick Watson. London: Pearson, 2003.

Crossan, John Dominic. *The Historical Jesus: The Life of a Mediterranean Jewish Peasant*. New York: HarperCollins, 1992.

———. "The Life of a Mediterranean Jewish Peasant." *Christian Century* 108 (1991): 1194-1200.

Crow, Liz. "Including All of Our Lives: Renewing the Social Model of Disability." In *Exploring the Divide: Illness and Disability*. Edited by Colin Barnes and Geof Mercer. Leeds: Disability Press, 1996.

Currat, Louis J. "The Global Health Situation in the 21st Century: Aspects from the Global Forum on Health Research and the World Health Organization in Geneva." *International Review of Mission* 95.376/377 (2006): 7-20.

Daniels, Norman. *Just Health: Meeting Health Needs Fairly*. Cambridge: Cambridge University Press, 2008.

———. *Just Health Care*. Cambridge: Cambridge University Press, 1985.

Darwin, Francis, ed. *The Life and Letters of Charles Darwin, including an Autobiographical Chapter*. 3 vols. London: John Murray, 1887.

Davis, Philip, and John Bradley. "The Meaning of Normal." *Perspectives in Biology and Medicine* 40.1 (1996): 68-77.

Deane-Drummond, Celia, and David Clough, eds. *Creaturely Theology: On God, Humans, and Other Animals*. London: SCM, 2009.

Deane-Drummond, Celia, and Peter Manley Scott, eds. *Future Perfect? God, Medicine, and Human Identity*. London: T&T Clark, 2006.
de Grey, Aubrey, with Michael Rae. *Ending Aging*. New York: St. Martin's Press, 2008.
de Grey, Aubrey D. N. J., Bruce N. Ames, Julie K. Andersen, Andrzej Bartke, Judith Campisi, Christopher B. Heward, Roger J. M. McCarter, and Gregory Stock. "Time to Talk SENS: Critiquing the Immutability of Human Aging." *Annals of the New York Academy of Sciences* 959 (2002): 452-462.
Delkeskamp-Hayes, Corinna. "Why Patients Should Give Thanks for Their Disease: Traditional Christianity on the Joy of Suffering." *Christian Bioethics* 12.2 (2006): 213-228.
Delkeskamp-Hayes, Corinna, ed. Theme Issue on "Sin and Disease." *Christian Bioethics* 12.2 (2006).
DeVito, Scott. "On the Value-Neutrality of the Concepts of Health and Disease: Unto the Breach Again." *Journal of Medicine and Philosophy* 25.5 (2000): 539-567.
de Vries, Christina. "Why Do Churches Need to Continue to Struggle for Health for All?" *International Review of Mission* 95.376/377 (2006): 21-35.
Disabled People's International. *Proceedings of the First World Congress, Singapore*. Singapore: Disabled People's International, 1982.
Dobzhansky, Theodosius. "Nothing in Biology Makes Sense Except in the Light of Evolution." *American Biology Teacher* 35 (1973): 125-129. Reprinted in *Evolution*. Edited by Mark Ridley. Oxford: Oxford University Press, 2004.
Duffy, Mary. "Making Choices." In *Mustn't Grumble: Writing by Disabled Women*. Edited by L. Keith. London: Women's Press, 1994.
Edgar, Andrew, Sam Salek, Darren Shickle, and David Cohen. *The Ethical QALY: Ethical Issues in Healthcare Resource Allocations*. Haslemere: Euromed Communications, 1998.
Eiesland, Nancy L. *The Disabled God: Towards a Liberatory Theology of Disability*. Nashville: Abingdon, 1994.
Engelhardt, H. Tristram, Jr. "The Disease of Masturbation: Values and the Concept of Disease." *Bulletin of the History of Medicine* 48 (1974): 234-248.
Evans, Abigail Rian. *Redeeming Marketplace Medicine: A Theology of Health Care*. Eugene, OR: Wipf and Stock, 2008.
Finkelstein, Vic. "The Commonality of Disability." in *Disabling Barriers — Enabling Environments*. Edited by John Swain, Vic Finkelstein, Sally French, and Mike Oliver. London: Sage, 1993.
———. "Outside, 'Inside Out.'" *Coalition* (April 1996): 30-36.
———. "A Personal Journey into Disability Politics." Paper presented at Centre for Disability Studies, University of Leeds, 2001. http://www.disability-archive.leeds.ac.uk/.
———. "The 'Social Model of Disability' and the Disability Movement." Disability Archive, March 2007. http://www.disability-archive.leeds.ac.uk/.

Ford, David. "On Being Theologically Hospitable to Jesus Christ: Hans Frei's Achievement." *Journal of Theological Studies*, NS 46.2 (1995): 532-546.
Forsyth, P. T. *The Justification of God: Lectures for War-time on a Christian Theodicy.* 2nd ed. London: Latimer House, 1948.
Francis of Assisi. "The Canticle of the Sun." Translated by Bill Barrett. http://www.webster.edu/~barrettb/canticle.htm.
Frei, Hans W. *Types of Christian Theology*. Edited by George W. Hunsinger and William C. Placher. New Haven: Yale University Press, 1992.
French, Sally. "Disability, Impairment, or Something in Between?" In *Disabling Barriers — Enabling Environments*. Edited by John Swain, Vic Finkelstein, Sally French, and Mike Oliver. London: Sage, 1993.
Fulford, K. W. M. *Moral Theory and Medical Practice*. Cambridge: Cambridge University Press, 1989.
———. "Praxis Makes Perfect: Illness as a Bridge Between Biological Concepts of Disease and Social Conceptions of Health." *Theoretical Medicine* 14 (1993): 305-320.
———. "Teleology without Tears: Naturalism, Neo-Naturalism, and Evaluationism in the Analysis of Function Statements in Biology (and a Bet on the Twenty-first Century)." *Philosophy, Psychiatry, and Psychology* 7.1 (2000): 77-94.
———. "'What Is (Mental) Disease?': An Open Letter to Christopher Boorse." *Journal of Medical Ethics* 27 (2001): 80-85.
Gaiser, Frederick J. *Healing in the Bible: Theological Insight for Christian Ministry*. Grand Rapids: Baker Academic, 2010.
Gallagher, Shaun. "Lived Body and Environment." *Research in Phenomenology* 16 (1986): 139-170.
Gammelgaard, Anne. "Evolutionary Biology and the Concept of Disease." *Medicine, Health Care, and Philosophy* 3 (2000): 109-116.
Gould, Stephen Jay. "Exaptation: A Crucial Tool for Evolutionary Analysis." *Journal of Social Issues* 47 (1991): 43-65.
Green, Joel B. *The Gospel of Luke*. New International Commentary on the New Testament. Grand Rapids: Eerdmans, 1997.
Green, John. *The Fault in Our Stars*. New York: Dutton, 2012.
———. "On Religion." http://www.youtube.com/watch?gl=GB&v=hXll8Wn8J3Q.
Gregory, Eric. "The Spirit and the Letter: Protestant Thomism and Nigel Biggar's 'Karl Barth's Ethics Revisited.'" In *Commanding Grace: Studies in Karl Barth's Ethics*. Edited by Daniel L. Migliore. Grand Rapids: Eerdmans, 2010.
Grundmann, Christoffer H. "Healing: A Challenge to Church and Theology." *International Review of Mission* 90.356/357 (2001): 26-40.
Gunton, Colin E. *Act and Being: Towards a Theology of the Divine Attributes*. London: SCM, 2002.
———. *The Actuality of Atonement: A Study of Metaphor, Rationality, and the Christian Tradition*. Edinburgh: T&T Clark, 1988.

Hacking, Ian. *The Taming of Chance*. Cambridge: Cambridge University Press, 1990.
Hare, Richard M. "Health." *Journal of Medical Ethics* 12 (1986): 174-181.
Harris, A., E. Cox, and C. Smith, *Handicapped and Impaired in Great Britain*, vol. 1 (London: HMSO, 1971).
Harris, John. *Clones, Genes, and Immortality: Ethics and the Genetic Revolution*. Oxford: Oxford University Press, 1998.
———. *Enhancing Evolution: The Ethical Case for Making Better People*. Princeton: Princeton University Press, 2007.
Hauerwas, Stanley. "Abled and Disabled: The Politics of Gentleness." *Christian Century* 125.24 (2008): 28-32.
———. "The Gesture of a Truthful Story." *Theology Today* 42.2 (1985): 181-189.
———. "The Radical Hope in the Annunciation: Why Both Single and Married Christians Welcome Children." In *The Hauerwas Reader*. Edited by John Berkman and Michael Cartwright. Durham, NC: Duke University Press, 2001.
———. *Suffering Presence: Theological Reflections on Medicine, the Mentally Handicapped, and the Church*. Edinburgh: T&T Clark, 1988.
Hauerwas, Stanley, with Richard Bondi and David B. Burrell. *Truthfulness and Tragedy: Further Investigations into Christian Ethics*. Notre Dame: University of Notre Dame Press, 1977.
Hauerwas, Stanley, and William H. Willimon. *Resident Aliens*. Nashville: Abingdon, 1989.
Hays, Richard B. *The Moral Vision of the New Testament: Community, Cross, New Creation*. Edinburgh: T&T Clark, 1997.
Hempel, Carl. "The Logic of Functional Analysis." In *Aspects of Scientific Explanation*. Edited by Carl Hempel. New York: Free Press, 1965.
Hick, John. *Evil and the God of Love*. London: Macmillan, 1966.
Hofmann, Bjørn. "On the Triad Disease, Illness, and Sickness." *Journal of Medicine and Philosophy* 27.6 (2002): 651-673.
Hope, Tony, K. W. M. Fulford, and Anne Yates. *The Oxford Practice Skills Course: Ethics, Law, and Communication Skills in Health Care Education*. Oxford: Oxford University Press, 1996.
Hume, David. *Dialogues Concerning Natural Religion*. London, 1779.
Humphrey, R. "Thoughts on Disability Arts." *Disability Arts Magazine* 4.1 (1994): 66-67.
Hurst, Rachel. "The International Disability Rights Movement and the ICF." *Disability and Rehabilitation* 25.11-12 (2003): 572-576.
———. "To Revise or Not to Revise?" *Disability and Society* 15.7 (2000): 1083-1087.
Iozzio, Mary Jo. "Genetic Anomaly or Genetic Diversity: Thinking in the Key of Disability on the Human Genome." *Theological Studies* 66 (2005): 862-881.
Iwakuma, Miho. "The Body as Embodiment: An Investigation of the Body by Merleau-Ponty." in *Disability/Postmodernity: Embodying Disability Theory*. Edited by Mairian Corker and Tom Shakespeare. London: Continuum, 2002.
Jefferys, M., J. B. Nullard, M. Hyman, and M. D. Warren. "A Set of Tests for Mea-

suring Motor Impairment in Prevalence Studies." *Journal of Chronic Diseases* 28 (1969): 303-309.
John Paul II. *Evangelium Vitae*. Vatican City, 1995. http://www.vatican.va/holy_father/john_paul_ii/encyclicals/documents/hf_jp-ii_enc_25031995_evangelium-vitae_en.html#-1M.
———. *Salvifici Doloris*. Vatican City, 1984. http://www.vatican.va/holy_father/john_paul_ii/apost_letters/documents/hf_jp-ii_apl_11021984_salvifici-doloris_en.html.
Johnson, Monte Ransome. *Aristotle on Teleology*. Oxford: Clarendon, 2005.
Kant, Immanuel. *Critique of the Power of Judgment*. Translated by Paul Guyer and Eric Matthews. Cambridge: Cambridge University Press, 2000.
Kaplan, A. *The Conduct of Inquiry*. San Francisco: Chandler, 1964.
Khushf, George. "An Agenda for Future Debate on Concepts of Health and Disease." *Medicine, Health Care, and Philosophy* 10 (2007): 19-27.
———. "Illness, the Problem of Evil, and the Analogical Structure of Healing: On the Difference Christianity Makes in Bioethics." *Christian Bioethics* 1.1 (1995): 102-120.
Kittay, Eva Feder. *Love's Labor: Essays on Women, Equality, and Dependency*. New York: Routledge, 1999.
Kovacs, Jozsef. "The Concept of Health and Disease." *Medicine, Health Care, and Philosophy* 1 (1998): 31-39.
Lammers, Stephen E., and Allen Verhey, eds. *On Moral Medicine: Theological Perspectives in Medical Ethics*. 2nd ed. Grand Rapids: Eerdmans, 1998.
Larchet, Jean-Claude. *The Theology of Illness*. Crestwood, NY: St. Vladimir's Seminary Press, 2002.
Law, Iain, and Heather Widdows. "Conceptualizing Health: Insights from the Capability Approach." *Health Care Analysis* 16 (2008): 303-314.
Leder, Drew. "Medicine and Paradigms of Embodiment." *Journal of Medicine and Philosophy* 9 (1984): 29-43.
Leder, Drew, ed. *The Body in Medical Thought and Practice*. Dordrecht: Kluwer Academic, 1992.
Lilienfeld, Scott O., and Lori Marino. "Mental Disorder as a Roschian Concept: A Critique of Wakefield's 'Harmful Dysfunction' Analysis." *Journal of Abnormal Psychology* 104 (1995): 411-420.
MacIntyre, Alasdair. *After Virtue: A Study in Moral Theory*. 2nd ed. London: Duckworth, 1985.
———. *Whose Justice? Which Rationality?* London: Duckworth, 1988.
MacKenzie, Catriona, and Jackie Leach Scully. "Moral Imagination, Disability, and Embodiment." *Journal of Applied Philosophy* 24.4 (2007): 335-351.
Marmot, Michael, and Richard G. Wilkinson, eds. *Social Determinants of Health*. 2nd ed. Oxford: Oxford University Press, 2006.
Mathey, Jacques. "Opening Biblical Reflection." *International Review of Mission* 95.376/377 (2006): 151-153.

Mathey, Jacques, ed. Theme Issue on "The Global Health Situation and the Mission of the Church in the Twenty-First Century." *International Review of Mission* 95.376/377 (2006).
McEwan, Ian. *Solar*. London: Vintage, 2011.
McFadyen, Alistair I. *Bound to Sin: Abuse, Holocaust, and the Doctrine of Sin*. Cambridge: Cambridge University Press, 2000.
McFarland, Ian A. "Who Is My Neighbor? The Good Samaritan as a Source for Theological Anthropology." *Modern Theology* 17.1 (2001): 57-66.
McKenny, Gerald P. *To Relieve the Human Condition: Bioethics, Technology, and the Body*. Albany: State University of New York Press, 1997.
McPherson, Kath. "What Are the Boundaries of Health and Functioning — and Who Should Say What They Are?" *Disability and Rehabilitation* 28.23 (2006): 1473-1474.
Megone, Christopher. "Aristotle's Function Argument and the Concept of Mental Illness." *Philosophy, Psychiatry, and Psychology* 5.3 (1998): 187-201.
———. "Mental Illness, Human Function, and Values." *Philosophy, Psychiatry, and Psychology* 7.1 (2000): 45-65.
Meilaender, Gilbert. *Bioethics: A Primer for Christians*. 3rd ed. Grand Rapids: Eerdmans, 2013.
Merleau-Ponty, Maurice. *The Phenomenology of Perception*. Translated by C. Smith. London: Routledge and Kegan Paul, 1962.
Merriam, Garret. "Rehabilitating Aristotle: A Virtue Ethics Approach to Disability and Human Flourishing." In *Philosophical Reflections on Disability*. Philosophy and Medicine 104. Edited by D. Christopher Ralston and Justin Ho. New York: Springer, 2010.
Messer, Neil. "The Human Genome Project, Health, and the 'Tyranny of Normality.'" In *Brave New World? Theology, Ethics, and the Human Genome*. Edited by Celia Deane-Drummond. London: T&T Clark, 2003.
———. "Natural Evil after Darwin." In *Theology after Darwin*. Edited by Michael Northcott and R. J. Berry. Carlisle: Paternoster, 2009.
———. *Respecting Life: Theology and Bioethics*. London: SCM, 2011.
———. *Selfish Genes and Christian Ethics: Theological and Ethical Reflections on Evolutionary Biology*. London: SCM, 2007.
———. "Toward a Theological Understanding of Health and Disease." *Journal of the Society of Christian Ethics* 31.1 (2011): 161-178.
Midgley, Mary. *Utopias, Dolphins, and Computers: Problems in Philosophical Plumbing*. London: Routledge, 1996.
Migliore, Daniel L. *Faith Seeking Understanding: An Introduction to Christian Theology*. Grand Rapids: Eerdmans, 1991.
Moltmann, Jürgen. *God in Creation: An Ecological Doctrine of Creation*. Translated by Margaret Kohl. London: SCM, 1985.
Morris, Jenny. *Pride against Prejudice: Transforming Attitudes to Disability*. London: The Women's Press, 1991.

Mumford, James. Review of Hans Reinders, *Receiving the Gift of Friendship. Studies in Christian Ethics* 23.2 (2010): 216-219.

Nagi, Saad Z. "Disability Concepts Revisited: Implications for Prevention." In *Disability in America: Toward a National Agenda for Prevention*. Edited by Andrew M. Pope and Alvin R. Tarlov. Washington, DC: National Academies Press, 1991.

National Institute for Health and Clinical Excellence (NICE). *Guide to the Methods of Technology Appraisal*. London: NICE, 2008.

Nichols, Terence L. "Miracles in Science and Theology." *Zygon* 37.3 (2002): 703-715.

Nordenfelt, Lennart. "Ability, Competence, and Qualification: Fundamental Concepts in the Philosophy of Disability." In *Philosophical Reflections on Disability*. Philosophy and Medicine 104. Edited by D. Christopher Ralston and Justin Ho. New York: Springer, 2010.

———. "The Concepts of Health and Illness Revisited." *Medicine, Health Care, and Philosophy* 10 (2007): 5-10.

———. "Concepts of Health and Their Consequences for Health Care." *Theoretical Medicine* 14 (1993): 277-285.

———. "Establishing a Middle-Range Position in the Theory of Health: A Reply to My Critics." *Medicine, Health Care, and Philosophy* 10 (2007): 29-32.

———. *Health, Science, and Ordinary Language*. Amsterdam: Rodopi, 2001.

———. "The Logic of Health Concepts." In *Handbook of Bioethics*. Edited by George Khushf. Dordrecht: Kluwer Academic, 2004.

———. "On Health, Ability, and Activity: Comments on Some Basic Notions in the ICF." *Disability and Rehabilitation* 28.23 (2006): 1461-1465.

———. *On the Nature of Health: An Action-Theoretic Approach*. 2nd ed. Dordrecht: Kluwer Academic, 1995.

———. "On the Relation between Biological and Social Theories of Health: A Commentary on Fulford's Praxis Makes Perfect." *Theoretical Medicine* 14 (1993): 321-324.

Nussbaum, Martha C. *Frontiers of Justice: Disability, Nationality, Species Membership*. Cambridge, MA: Harvard University Press, 2006.

———. "Nature, Function, and Capability: Aristotle on Political Distribution." *Oxford Studies in Ancient Philosophy* suppl. vol. (1988).

———. "Non-Relative Virtues: An Aristotelian Approach." In *The Quality of Life*. Edited by Martha C. Nussbaum and Amartya Sen. Oxford: Clarendon, 1993.

———. *Women and Human Development*. Cambridge: Cambridge University Press, 2000.

———. "Women's Capabilities and Social Justice." *Journal of Human Development* 1.2 (2000): 219-229.

Oliver, Michael. "Defining Impairment and Disability: Issues at Stake." In *Exploring the Divide: Illness and Disability*. Edited by Colin Barnes and Geof Mercer. Leeds: Disability Press, 1996.

———. *The Politics of Disablement*. London: Macmillan, 1990.

———. "The Social Model in Action: If I Had a Hammer." In *Implementing the Social Model of Disability: Theory and Research*. Edited by Colin Barnes and Geof Mercer. Leeds: Disability Press, 2004.

———. *Social Work with Disabled People*. Basingstoke: Macmillan, 1983.

———. *Understanding Disability: From Theory to Practice*. 2nd ed. Basingstoke: Palgrave Macmillan, 2009.

Oliver, Mike, and Colin Barnes. "Disability Politics and the Disability Movement in Britain: Where Did It All Go Wrong?" Disability Archive, June 2006. http://www.disability-archive.leeds.ac.uk/.

Oliver, Mike, and Len Barton. "The Emerging Field of Disability Studies: A View from Britain." Paper presented at Disability Studies: A Global Perspective, Washington, DC, October 2000. http://www.disability-archive.leeds.ac.uk/.

Oliver, Simon. "Analytic Theology: A Review Article on Oliver D. Crisp and Michael C. Rea, eds., *Analytic Theology: New Essays in the Philosophy of Theology*." *International Journal of Systematic Theology* 12.4 (2010): 464-475.

———. "Aquinas and Aristotle's Teleology." *Nova et Vetera* 11.3 (forthcoming, 2013).

Oregon Public Health Division. "Oregon's Death with Dignity Act — 2011." Salem: State of Oregon, 2011. http://public.health.oregon.gov/ProviderPartnerResources/EvaluationResearch/DeathwithDignityAct/Documents/year14.pdf.

Patrick, D., and H. Peach, eds. *Disablement in the Community*. Oxford: Oxford Medical Publications, 1989.

Peacocke, Arthur. "Biology and a Theology of Evolution." *Zygon* 34.4 (1999): 695-712.

Perry, John, ed. *A Time to Heal: A Report for the House of Bishops on the Healing Ministry*. London: Church House Publishing, 2000.

Peters, Ted. *Playing God? Genetic Determinism and Human Freedom*. New York: Routledge, 1997.

Peterson, Gregory R. "Are Evolutionary/Cognitive Theories of Religion Relevant for Philosophy of Religion?" *Zygon* 45.3 (2010): 545-557.

Pfeiffer, David. "The Devils Are in the Details: The ICIDH2 and the Disability Movement." *Disability and Society* 15.7 (2000): 1079-1082.

———. "The ICIDH and the Need for Its Revision." *Disability and Society* 13.4 (1998): 503-523.

Phillips, D. Z. *The Problem of Evil and the Problem of God*. London: SCM, 2004.

Pinder, Ruth. "Sick-but-Fit or Fit-but-Sick? Ambiguity and Identity at the Workplace." In *Exploring the Divide: Illness and Disability*. Edited by Colin Barnes and Geof Mercer. Leeds: Disability Press, 1996.

Pope, Andrew M., and Alvin R. Tarlov. *Disability in America: Toward a National Agenda for Prevention*. Washington, DC: National Academies Press, 1991.

Pope, Stephen J. *Human Evolution and Christian Ethics*. Cambridge: Cambridge University Press, 2007.

Porter, Jean. *The Recovery of Virtue: The Relevance of Aquinas for Christian Ethics.* London: SPCK, 1994.

Price, Janet, and Margrit Shildrick. "Bodies Together: Touch, Ethics, and Disability." In *Disability/Postmodernity: Embodying Disability Theory.* Edited by Mairian Corker and Tom Shakespeare. London: Continuum, 2002.

Ramsey, Paul. *The Patient as Person: Explorations in Medical Ethics.* New Haven: Yale University Press, 1970.

Ratcliffe, Matthew. *Feelings of Being: Phenomenology, Psychiatry, and the Sense of Reality.* Oxford: Oxford University Press, 2008.

Ratcliffe, Matthew, Matthew Broome, Benedict Smith, and Hannah Bowden. "A Bad Case of the Flu? The Comparative Phenomenology of Depression and Somatic Illness." *Journal of Consciousness Studies* (forthcoming, 2013).

Rawls, John. *Political Liberalism.* Expanded ed. New York: Columbia University Press, 2005.

———. *A Theory of Justice.* Cambridge, MA: Harvard University Press, 1971.

Reed, John Shelton. "'Ritualism Rampant in East London': Anglo-Catholicism and the Urban Poor." *Victorian Studies* 31.3 (1988): 375-403.

Reinders, Hans S. *Receiving the Gift of Friendship: Profound Disability, Theological Anthropology, and Ethics.* Grand Rapids: Eerdmans, 2008.

———. "Understanding Humanity and Disability: Probing an Ecological Perspective." *Studies in Christian Ethics* 26.1 (2013): 37-49.

Reiss, Michael J. "And in the World to Come, Life Everlasting." In *Brave New World? Theology, Ethics, and the Human Genome.* Edited by Celia Deane-Drummond. London: T&T Clark, 2003.

Reynolds, Thomas E. *Vulnerable Communion: A Theology of Disability and Hospitality.* Grand Rapids: Brazos, 2008.

Rogers, Eugene F., Jr. *Thomas Aquinas and Karl Barth: Sacred Doctrine and the Natural Knowledge of God.* Notre Dame: University of Notre Dame Press, 1995.

Romero, Miguel J. "Aquinas on the *Corporis Infirmitas:* Broken Flesh and the Grammar of Grace." In *Disability in the Christian Tradition: A Reader.* Edited by Brian Brock and John Swinton. Grand Rapids: Eerdmans, 2012.

Rosch, Eleanor R. "Natural Categories." *Cognitive Psychology* 4 (1973): 328-350.

Rose, Steven. *The Future of the Brain.* Oxford: Oxford University Press, 2005.

Ruse, Michael. *Darwin and Design: Does Evolution Have a Purpose?* Cambridge, MA: Harvard University Press, 2003.

Rutherford, Adam. "Alpha Can't Heal My Skepticism." *The Guardian Online,* 4 September 2009. http://www.guardian.co.uk/commentisfree/belief/2009/sep/03/alpha-course-religion-christianity.

Sacks, Oliver. *A Leg to Stand On.* New York: Summit, 1984.

———. *The Man Who Mistook His Wife for a Hat and Other Clinical Tales.* New York: Summit, 1985.

———. *The Mind's Eye.* London: Picador, 2011.

Sartre, Jean-Paul. *Being and Nothingness: An Essay on Phenomenological Ontology.* London: Methuen, 1957.

Schneidert, Marguerite, Rachel Hurst, Janice Miller, and Bedirhan Üstün. "The Role of Environment in the International Classification of Functioning, Disability, and Health (ICF)." *Disability and Rehabilitation* 25.11-12 (2003): 588-595.

Schramme, Thomas. "Lennart Nordenfelt's Theory of Health: Introduction to the Theme." *Medicine, Health Care, and Philosophy* 10 (2007): 3-4.

———. "A Qualified Defense of a Naturalist Theory of Health." *Medicine, Health Care, and Philosophy* 10 (2007): 11-17.

Scully, Jackie Leach. "A Postmodern Disorder: Moral Encounters with Molecular Models of Disability." In *Disability/Postmodernity: Embodying Disability Theory.* Edited by Mairian Corker and Tom Shakespeare. London: Continuum, 2002.

Sedgwick, Peter. *Psycho Politics.* New York: Harper and Row, 1982.

Seedhouse, David. *Health: Foundations of Achievement.* Chichester: Wiley, 1986.

Sen, Amartya. "Capability and Well-Being." In *The Quality of Life.* Edited by Martha C. Nussbaum and Amartya Sen. Oxford: Clarendon, 1993.

———. "Equality of What?" Tanner Lecture on Human Values. Stanford University, 22 May 1979. http://www.uv.es/~mperezs/intpoleco/Lecturcomp/Distribucion%20Crecimiento/Sen%20Equaliy%20of%20what.pdf.

———. *Inequality Reexamined.* Cambridge, MA: Harvard University Press, 1992.

Shakespeare, Tom. "Disability: Suffering, Social Oppression, or Complex Predicament?" In *The Contingent Nature of Life: Bioethics and the Limits of Human Existence.* Edited by Marcus Düwell, Christoph Rehmann-Sutter, and Dietmar Mieth. New York: Springer, 2008.

———. *Disability Rights and Wrongs.* London: Routledge, 2006.

Shakespeare, Tom, and Nicholas Watson. "The Social Model of Disability: An Outdated Ideology?" *Research in Social Science and Disability* 2 (2002): 9-28.

Shelton, James B. "'Not Like It Used to Be'? Jesus, Miracles, and Today." *Journal of Pentecostal Theology* 14.2 (2006): 219-227.

———. "A Reply to Keith Warrington's Response to 'Jesus and Healing: Yesterday and Today.'" *Journal of Pentecostal Theology* 16 (2008): 113-117.

Shildrick, Margrit, and Janet Price. "Breaking the Boundaries of the Broken Body." *Body and Society* 2.4 (1996): 93-113.

Shuman, Joel, and Brian Volck. *Reclaiming the Body: Christians and the Faithful Use of Modern Medicine.* Grand Rapids: Brazos, 2006.

Silvers, Anita. "An Essay on Modeling: The Social Model of Disability." In *Philosophical Reflections on Disability.* Philosophy and Medicine 104. Edited by D. Christopher Ralston and Justin Ho. Dordrecht: Springer, 2010.

———. "A Fatal Attraction to Normalizing: Treating Disabilities as Deviations from 'Species-Typical' Functioning." In *Enhancing Human Traits: Ethical and Social Implications.* Edited by Erik Parens. Washington, DC: Georgetown University Press, 1998.

Simmons, Fred. "Contemporary Cosmologies and Christian Faith." Journey of the

Universe, 2012. http://www.journeyoftheuniverse.org/storage/Fred.Simmons.pdf.
Slovo, Gillian. *Every Secret Thing: My Family, My Country*. London: Little, Brown, 1997.
Song, Robert. *Human Genetics: Fabricating the Future*. London: Darton, Longman, and Todd, 2002.
Southgate, Christopher. *The Groaning of Creation: God, Evolution and the Problem of Evil*. Louisville: Westminster John Knox, 2008.
———. "Re-reading Genesis, John, and Job: a Christian Response to Darwinism." *Zygon* 46.2 (2011): 370-395.
Sparrow, Robert. "Defending Deaf Culture: The Case of Cochlear Implants." *Journal of Political Philosophy* 13.2 (2005): 135-152.
Stahl, Elisabeth, et al. "Health-Related Quality of Life, Utility, and Productivity Outcomes Instruments: Ease of Completion by Subjects with COPD." *Health and Quality of Life Outcomes* 1 (2003): 18.
Sulmasy, Daniel P. "Dignity, Disability, Difference, and Rights." In *Philosophical Reflections on Disability*. Philosophy and Medicine 104. Edited by D. Christopher Ralston and Justin Ho. New York: Springer, 2010.
Swinton, John. "Known by God." In *The Paradox of Disability: Responses to Jean Vanier and the L'Arche Community from Theology and the Sciences*. Edited by Hans S. Reinders. Grand Rapids: Eerdmans, 2010.
———. *Raging with Compassion: Pastoral Responses to the Problem of Evil*. Grand Rapids: Eerdmans, 2007.
Szasz, Thomas S. "Mental Illness Is Still a Myth." *Society* 31.4 (1994): 34-39.
———. *The Myth of Mental Illness: Foundations of a Theory of Personal Conduct*. Revised ed. New York: Harper and Row, 1974.
———. "Second Commentary on 'Aristotle's Function Argument.'" *Philosophy, Psychiatry, and Psychology* 7.1 (2000): 3-16.
Thomas Aquinas. *Summa Contra Gentiles*. Edited by Joseph Kenny. Translated by Vernon J. Bourke. New York: Hanover House, 1955-57.
———. *Summa Theologica*. Translated by Fathers of the English Dominican Province. 1920. http://www.newadvent.org/summa/index.html.
Thomas, Carol. "Developing the Social Relational in the Social Model of Disability: a Theoretical Agenda." In *Implementing the Social Model of Disability: Theory and Research*. Edited by Colin Barnes and Geof Mercer. Leeds: The Disability Press, 2004.
Thomas, Carol, and Mairian Corker. "A Journey around the Social Model." In *Disability/Postmodernity: Embodying Disability Theory*. Edited by Mairian Corker and Tom Shakespeare. London: Continuum, 2002.
Thrall, Margaret E. *A Critical and Exegetical Commentary on the Second Epistle to the Corinthians*, vol. 2. Edinburgh: T&T Clark, 2000.
Toombs, S. Kay. *The Meaning of Illness: A Phenomenological Account of the Different Perspectives of Physician and Patient*. Dordrecht: Kluwer Academic, 1992.

Topliss, Eda. *Provision for the Disabled.* 2nd ed. Oxford: Blackwell, 1979.
Tremain, Shelley. "On the Subject of Impairment." In *Disability/Postmodernity: Embodying Disability Theory.* Edited by Mairian Corker and Tom Shakespeare. London: Continuum, 2002.
Ueda, S., and Y. Okawa. "The Subjective Dimension of Functioning and Disability: What Is It and What Is It For?" *Disability and Rehabilitation* 25.11-12 (2003): 596-601.
Union of the Physically Impaired Against Segregation. *Fundamental Principles of Disability.* London: Union of the Physically Impaired Against Segregation, 1976.
Üstün, T. B., S. Chatterji, J. Bickenbach, N. Kostanjsek, and M. Schneider. "The International Classification of Functioning, Disability, and Health: A New Tool for Understanding Disability and Health." *Disability and Rehabilitation* 25.11-12 (2003): 565-571.
Verkerk, M. A., J. J. V. Busschbach, and E. D. Karssing. "Health-Related Quality of Life Research and the Capability Approach of Amartya Sen." *Quality of Life Research* 10 (2001): 49-55.
Volpe, Medi Ann. "Irresponsible Love: Rethinking Intellectual Disability, Humanity, and the Church." *Modern Theology* 25.3 (2009): 491-501.
von Uexküll, Jakob. "A Stroll Through the Worlds of Animals and Men." In *Instinctive Behavior: The Development of a Modern Concept.* Edited and translated by Claire H. Schiller. New York: International Universities Press, 1957.
Wakefield, Jerome C. "Aristotle as Sociobiologist: The 'Function of a Human Being' Argument, Black Box Essentialism, and the Concept of Mental Disorder." *Philosophy, Psychiatry, and Psychology* 7.1 (2000): 17-44.
———. "The Concept of Mental Disorder: On the Boundary Between Biological Facts and Social Values." *American Psychologist* 47.3 (1992): 373-388.
———. "Evolutionary versus Prototype Analyses of the Concept of Disorder." *Journal of Abnormal Psychology* 108.3 (1999): 374-399.
Walmsley, Jan. "Including People with Learning Difficulties: Theory and Practice." In *Disability Studies: Past, Present, and Future.* Edited by Len Barton and Mike Oliver. Leeds: Disability Press, 1997.
Walters, LeRoy. "Paul Braune Confronts the National Socialists' 'Euthanasia' Program." *Holocaust and Genocide Studies* 21.3 (2007): 454-487.
Wannenwetsch, Bernd. "From *Ars Moriendi* to Assisted Suicide: Bonhoefferian Explorations into Cultures of Death and Dying." *Studies in Christian Ethics* 24.4 (2011): 428-440.
Warrington, Keith. "Acts and the Healing Narratives: Why?" *Journal of Pentecostal Theology* 14.2 (2006): 189-217.
———. "James 5:14-18: Healing Then and Now." *International Review of Mission* 93.370/371 (2004): 346-367.
———. "Healing and Suffering in the Bible." *International Review of Mission* 95.376/377 (2006): 154-164.

———. "The Path to Wholeness: Beliefs and Practices Relating to Healing in Pentecostalism." *Evangel* 21.2 (2003): 45-49.

———. "A Response to James Shelton concerning Jesus and Healing: Yesterday and Today." *Journal of Pentecostal Theology* 15.2 (2007): 185-193.

Wasserman, David. "Philosophical Issues in the Definition and Social Response to Disability." In *Handbook of Disability Studies*. Edited by Gary L. Albrecht, Katherine D. Seelman, and Michael Bury. Thousand Oaks, CA: Sage, 2001.

Wasserman, David, Adrienne Asch, Jeffrey Blustein, and Daniel Putnam. "Disability: Definitions, Models, Experience." In *The Stanford Encyclopedia of Philosophy* (Winter 2011 edition). Edited by Edward N. Zalta. Stanford: Stanford University Press, 2011. http://plato.stanford.edu/archives/win2011/entries/disability/.

Waters, Brent. "Disability and the Quest for Perfection: A Moral and Theological Enquiry." In *Theology, Disability, and the New Genetics: Why Science Needs the Church*. Edited by John Swinton and Brian Brock. London: T&T Clark, 2007.

———. *From Human to Posthuman: Christian Theology and Technology in a Postmodern World*. Aldershot: Ashgate, 2006.

Waters, Brent, and Ronald Cole-Turner, eds. *God and the Embryo: Religious Voices on Stem Cells and Cloning*. Washington, DC: Georgetown University Press, 2003.

Wells, Samuel. "The Disarming Virtue of Stanley Hauerwas." *Scottish Journal of Theology* 52.1 (1999): 82-88.

Wilkinson, John. *The Bible and Healing: A Medical and Theological Commentary*. Grand Rapids: Eerdmans, 1998.

Williams, Alan. "Economics of Coronary Artery Bypass Grafting." *British Medical Journal* 291 (1985): 326-329.

Williams, Gareth. "Theorizing Disability." In *Handbook of Disability Studies*. Edited by Gary L. Albrecht, Katherine D. Seelman, and Michael Bury. Thousand Oaks, CA: Sage, 2001.

Wilson, David Sloan. *Darwin's Cathedral: Evolution, Religion, and the Nature of Society*. Chicago: University of Chicago Press, 2003.

World Council of Churches Christian Medical Commission. *Healing and Wholeness*. Geneva: World Council of Churches, 1990.

World Health Organization. *Basic Documents*. 46th ed. Geneva: World Health Organization, 2007. http://www.who.int/gb/bd/.

———. *International Classification of Functioning, Disability, and Health*. Short Version. Geneva: World Health Organization, 2001.

———. *International Classification of Impairments, Disabilities, and Handicaps*. Geneva: World Health Organization, 1980.

Yong, Amos. *Theology and Down Syndrome: Reimagining Disability in Late Modernity*. Waco: Baylor University Press, 2007.

Zaner, Richard M. *The Context of Self: A Phenomenological Inquiry Using Medicine as a Clue*. Athens, OH: Ohio University Press, 1981.

Index

Abberley, Paul, 58-59, 63-64
Admirand, Peter, 109n
Albrecht, Gary L., 23n, 52n, 71n, 86n, 179n, 209n
Almeder, Robert F., 7n
Altman, Barbara M., 51n, 54, 56-57
Altruistic behavior, 28n
American Psychiatric Association's *Diagnostic and Statistical Manual (DSM)*, xiii, 37
Americans with Disabilities Act (ADA) (1990), 58
Amundson, Ron, 8n, 85n, 209n; on concept of normality in disability definitions, 69-70, 80-82, 97-98, 100, 178, 190, 206; criticism of socio-medical model of disability, 69-70, 71, 86, 92; and natural kind concepts, 95, 97-98
Ananth, Mahesh, 27n
Aristotle: concept of *eudaimonia*, 14, 39-40, 170; distinction between intrinsic and extrinsic teleologies, 145-46; distinction between primary and secondary substances, 144n, 145; *Ethics*, 39; fourfold scheme of causation, 44-46, 145; function argument, 41-43, 47-48, 98-99; and Megone's teleological account of natural functions, 38-48; Merriam's exploration of virtue ethics, 93-95, 167, 172-74; *Nicomachean Ethics*, 39; notion of rationality as function, 41-42, 98-99; and Thomistic teleology, 144-47, 165-66. *See also* Teleological accounts of health
Arneson, Richard, xii
Arnhart, Larry, 181n
Arras, John D., 4n
Asch, Adrienne, 52n

Bacon, Francis, 47n, 124-26, 194, 198
Baconian project, McKenny's critique of, 124-26, 198
Barnes, Colin, 52n, 54n, 59n, 62n, 189n
Barrett, C. K., 120n, 135
Barth, Karl, 105-6, 134-41; affirmation of medical doctors' roles, 138; attempted *rapprochements* between Aquinas and, 141-44, 165-66; and Christian ambivalence about death, 194, 195; critique of Thomistic ethics, 141-44, 165-66; on disease, illness, and sickness, 139-41, 160, 184-88, 192-93; on evil as nothingness, 139-40, 186, 191-92; on health and ethics of creation, xv-xvi, 134-41, 164-66, 171-74; on health as a capability, 25n, 138, 176-79; on health as a penultimate good, 140-41, 160, 181-82; on health as "strength for human life," 49, 136, 138-39, 155, 175-79, 184, 202, 207; on health as wholeness, 138, 175,

227

176; and human flourishing, 105-6, 136-41, 175-79; on limits of the natural sciences, 180-81; on social and political dimension to health, 139, 182-84, 208; on "the will to be healthy," 136-37, 176, 182-84, 186, 197, 202, 209
Barton, Len, 61n, 66n
Basic vital goals: Fulford on Nordenfelt's concept of, 16-18; and Nordenfelt's holistic theory of health, 12, 13-14, 16-18, 46-47, 170
Basnett, Ian, 71n, 85, 179, 190n, 209n
Bayer, Ronald, xiii
Beasley-Murray, George R., 129n
Begum, Nasa, 65n
Berkman, John, 137n
Berry, R. J., 131n, 194n
Bickenbach, J., 67n
Biggar, Nigel, 136n, 143-44, 174n, 193
"Biopower," 23, 125, 164
Biostatistical model of health (Boorse), xii, 6-10, 15, 18, 32, 79-80, 94, 121n; concept of hierarchy, 7; concept of species-typical natural functions, 8-9, 79-80; contrasted to WHO definition, 6-10; criticisms, 8-10, 15, 18, 32, 79-80, 206; and Daniels's resource allocation theory, xii, 206; disease/illness distinction, 6-7; and evolutionary biology, 7, 8-9; fact-value separation, 10, 15, 18; and Fulford's reverse theory, 16-18; limits of concept of health, 7; and Nordenfelt's holistic theory, 13, 15; and social model of disability, 79-80n; value-laden language, 6-10, 15, 18, 32
Blustein, Jeffrey, 52n
Bondi, Richard, xvii
Bonhoeffer, Dietrich, xvii; on penultimate and ultimate goods, xvii, 140-41, 170, 181; poem "Stations on the Road to Freedom" (on suffering), 133n; reading of the Genesis 3 Fall narrative, 195; term "human flourishing," 12n
Book of Common Prayer (Anglican), 114-15, 196n
Boorse, Christopher, xii, 6-10, 15, 18, 32, 79-80, 82-83, 94, 121n, 206. See also Biostatistical model of health (Boorse)

Boyle, Joseph, 5n
Braddock, David L., 57-58, 66n
Bradley, John, 69n
British social model. See Social model of disability
Brock, Brian, 150n, 158n, 173
Brock, Stephanie, 158n
Brody, Baruch, 95
Bromiley, Geoffrey W., 25n
Buller, Cornelius A., 112n
Burrell, David B., xvii
Bury, Michael, 23n, 52n, 54n, 60, 62, 67, 70, 179n, 209n
Busschbach, J. J. V., 24n

Callahan, Daniel, 2-4, 118n
Campbell, Alistair, 47
Capabilities approach to health, 14, 15, 23-26, 89-92, 176-79, 184; advantages of, 25; Barth on health as a capability, 25n, 138, 176-79; deliberate incompleteness and pluralistic character of, 25-26, 177; and disability, 89-92, 178-79; Law and Widdows, 14, 15, 23-26, 89, 176-77, 184; Nussbaum, xii, 14, 23, 26, 89-92, 176-79; Sen, xii, 2, 14, 23-26, 89, 176-77; ten basic capabilities (Nussbaum), 90-91
Cartwright, Michael, 137n
Caspary, Almut, xi, 107n
Cassell, Eric J., 23n
Chapman, Philip, 166n
Chappell, Anne Louise, 66n
Charlton, James I., 53n
Chatterji, S., 67n
Cherry, Mark J., 5n
Chiquete, Daniel, 111n, 116
Chisholm, Brock, 3
Clayton, Philip, 181n
Clements, Keith, 133n
Clough, David, 94n
Cochrane, James R., 138n
Cognitive impairments and intellectual disabilities: and British social model of disability, 66; Down syndrome, 84, 91, 98-99, 151-52, 157-58, 169, 186, 205; Hauerwas and Willimon on Christian communities and, 151-52, 157, 174, 186,

205; recent theological reflections on, 151-60, 169, 186, 205; Volpe on encounters with, 154, 155; Yong on social constructions, 157, 169. *See also* Mental disorders
Cohen, David, 85n
Cole-Turner, Ronald, 205n
Common good, concept of, 207
Corker, Mairian, 63n, 66, 100n
Cox, E., 67n
Cranmer, Thomas, 115, 196
Creatures, humans as. *See* Humans as creatures
Crossan, John Dominic, 116, 118-19
Crow, Liz, 60-61, 189n
Currat, Louis J., 183n

Daniels, Norman: and Boorse's biostatistical theory, xii, 206; health care resource allocation theory, xii-xiii, 206-8; *Just Health*, 207; and Rawls's theory of justice, xii, 207-8
Darwin, Charles, 94, 127-28n, 167n. *See also* Evolutionary models of health, disease, and illness
Davis, Philip, 69n
De Grey, Aubrey, 182n, 194n, 196, 204
De Vries, Christina, 183n
Deafness, 8, 53, 82, 152, 204; children and, 70, 100, 205; congenital, 82, 84; Deaf community activists, 53, 100, 205
Deane-Drummond, Celia, 5n, 94n, 134n, 141n, 173n, 182n
Death, Christian ambivalence about, 194-97, 203
Delkeskamp-Hayes, Corinna, 188n, 192n, 193n
DeVito, Scott, 8, 10, 15, 31-32
Devlieger, Patrick J., 86n
Diagnostic and Statistical Manual of Mental Disorders (DSM), xiii, 37
Disability Alliance, 55
Disability Discrimination Act (1995) (UK), 59
Disability models, 54-78; and cognitive impairments, 66; and concept of normality, 69-70, 73-74, 79, 190; differences and definitions, 54-55, 62; and disability rights movement/organizations, 53, 55-60, 63, 66, 68-71, 72; feminist critiques, 65-66; and medicalization of disability, 56, 59-62, 69; "minority-group" model (U.S.), 55, 57-59; QoL assumptions, 70-71, 74; social model (British), 54, 55-66, 99, 189-90; socio-medical models, 54, 67-78; terminology in discussing disability, 52-53, 71, 76; WHO's *ICF*, xv, 71-78, 83-84, 190; WHO's *ICIDH* definitions, 67-71, 79. *See also* Disability perspectives
"Disability paradox," 86n
Disability perspectives, xv, 51-101, 163-64; and Boorse's biostatistical model, 80, 82-83; capabilities approach (Nussbaum), 89-92, 178-79; and concept of normality, 69-70, 73-74, 79-82, 164, 183, 190; critical insights for theological account of health, 99-101, 163-64; disability/impairment distinctions, 56, 59-60, 61, 63-65, 189-90; health and disability, 78-99; Merriam's exploration of Aristotelian virtue ethics, 93-95, 167, 172-74; and notion of natural kinds, 93-94, 95-98, 167; phenomenological perspectives, 87-89; QoL assessments, xiii, 70-71, 74, 84-86, 208-9; recent theological literature on disability, 14n, 151-60, 169; and the scope/limits of "health" and "disability," 82-84; and teleological accounts of health, 92-99, 101; terminology issues, 52-53, 71, 76. *See also* Disability models; Theological literature on disability
Disability-adjusted life years (DALYs), 74, 85
Disabled People's International (DPI), 56, 62, 65
Disease/illness distinctions, 1n, 6-7, 15-16, 20, 191-93, 203
Disease/impairment distinctions, 62-63, 78-79, 189-90
Disorder: Lilienfeld and Marino's prototype model, 36-38, 185; Wakefield's "harmful dysfunction" model, 27-28,

31-33, 37-38, 179-80. *See also* Mental disorders
Dobzhansky, Theodosius, 32
Down syndrome, 84, 91, 98-99, 151-52, 157-58, 169, 186, 205
Draper, W. H., 195-96
"Drapetomania," 15
Duffy, Mary, 70n
Düwell, Marcus, 77n

Edgar, Andrew, 85n
Education for All Handicapped Children Act (1975), 58
Eisland, Nancy L., 129n
Engelhardt, H. Tristram, Jr., 5n, 15n, 37
Eudaimonia, 14, 39-40, 170
Evans, Abigail Rian, 108n, 113, 114n, 115-16, 118n, 119-21, 175n
Evil, suffering, and sin (disease and), xvii, 46n, 50, 126-34, 130n, 184-97, 202-3; Barth on evil as nothingness, 139-40, 186, 191-92; Book of Job, 129; and Christian ambivalence about death, 194-97, 203; complex connections between disease and evil/sin, 130n, 186-89, 191-93, 203; complex relationship between disease, impairment, and disability, 189-90, 203; disease as threats to creaturely flourishing, 184-86, 202-3; disease/illness distinctions, 191-93, 203; Green's *The Fault in Our Stars*, 126-27, 128, 132-33; and the healing ministry of Jesus, 128-30, 188; John's account of the healing of the man born blind, 129-30, 131, 187; loss of dignity in terminally ill patients, 132; philosophical theodicies, 130-31; practical theodicies, 131-34; theodicy in the context of Christian healing ministry, 126-34, 188; and Thomistic teleology, 150
Evolutionary models of health, disease, and illness, 28-38, 41-49, 79-82; and adaptation, 33-35, 81-82; Amundson's critiques, 81-82, 180, 185; and Aristotelian teleology, 29, 41-48, 146; and Boorse's biostatistical model, 7, 8-9, 79-80; evolutionary basis of natural

functions, 8-9, 28-38, 41-43, 49, 79-82, 179-80; and exaptations, 33-35; inclusive fitness, 28; Lilienfeld and Marino's critique of Wakefield's concepts, 33-38, 81n; Lilienfeld and Marino's prototype model (Roschian concepts), 36-38, 185; and Megone's teleological account of health as natural function, 10, 38-48, 92-93, 137, 191; and mental functions/mental disorders, 27, 28n, 33-38; and suffering associated with disease, 127; and theology, 29n; Wakefield/Megone dispute over nature of teleological explanation, 41-48, 168, 171, 180; Wakefield's "harmful dysfunction" model of disease and disorder, 26-38, 100, 179-80, 185

The Fault in Our Stars (Green), 126-27, 128, 132-33
Fenton, Elizabeth M., 4n
Finkelstein, Vic, 52n, 55n, 61n, 62, 66n
Flourishing, human. *See* Human flourishing
Ford, David, 103n
Forsyth, P. T., 128n
Foucault, Michel: analysis of "biopower," 23, 125, 164; and social view of impairment, 64-65
Francis of Assisi, 195-96
Frei, Hans W., 103n, 104
French, Sally, 52n, 61
Fulford, K. W. M. (Bill), 12n; on Boorse's biostatistical model of health, 6, 9-10, 16-18; bridge theory (illness as conceptual priority), 10-11, 16-18; disease/illness distinction, 20, 191; on Nordenfelt's concept of basic vital goals, 16-18

Gaiser, Frederick J., 106, 113-14, 116n, 119, 122, 124, 195
Gallagher, Shaun, 21
Gammelgaard, Anne, 8
Gellert, C. F., 196n
Genetic therapy/enhancement distinction, xi-xii, 204-6
"God of the gaps" views, 123-24

Gould, Stephen Jay, 33
Green, Joel B., 113n
Green, John, 126-27, 128, 132-33
Gregory, Eric, 143n
Grove-White, Robin, 173n
Grundmann, Christoffer H., 111n, 114
Gunton, Colin E., 128n, 159n, 182

Hacking, Ian, 52
Hamilton, William, 28n
Hare, Richard M., 10n
"Harmful dysfunction" model of disease and disorder (Wakefield), xiv, 26-36, 100, 179-80, 185; and adaptation, 33-35, 81-82; Amundson's critique, 81-82, 180, 185; and Aristotelian teleology, 29, 41-48, 146, 168, 171, 180; and Boorse's biostatistical theory, 27n, 31; and concept of inclusive fitness, 28; criterion of harm, 32-33; DeVito's critique, 31-32; and evolutionary basis of natural functions, 8-9, 28-38, 41-43, 49, 81-82, 179-80; and exaptations, 33-35; fact-value distinctions, 31-33, 37-38, 44, 46; Lilienfeld and Marino's critiques, 33-38, 81n; and meaning of suffering, 127; Megone-Wakefield dispute over nature of teleological explanation (and relationship between natural function and the good), 41-48, 168, 171, 180; and mental functions, 27, 28n, 33-36; term "disorder," 27-28, 31-33, 37-38, 179-80; value-laden language, 31-32, 37-38, 180; Wakefield's responses to critics, 34-35, 38n
Harris, A., 67n
Harris, John, xii, 86
Hauerwas, Stanley: on Christian ambivalence about death, 196-97; on distinctive practices of the Church community, xvii, 106, 125n, 134, 174; on healing ministries when cure isn't possible, xvii, 199-200; on medicine as a "tragic profession," 131-32; story of Dorothy (theological perspective on disability), 151, 157, 186, 205
Hays, Richard B., 169n
Healing ministries, Christian, xv, 107-34; charismatic aspects, 109-10; and Church of England report (*A Time to Heal*), 109, 110, 112, 116-17, 122-23, 130-31; curing disease–healing illness distinctions, 116-21; evangelism and Christian witness, 110-12; and forgiveness of sins, 114; and human flourishing, 117; Jesus' healing ministry in the Gospels and James, 107-10, 128-30, 188; John's account of the healing of the man born blind, 129-30, 131, 187; and the medical model/scientific medicine, 115-26, 197-99, 203; and New Testament vocabulary of healing, 113-14; pastoral character, 109; and Pentecostalism, 108n, 110, 111, 116, 121-22; as restoration of relationships, 113-14; sacramental character, 109; and supernatural/miraculous healings, 123, 123-24n; tasks when cure is not possible, xvii, 199-200, 203; theodicy in context of (and Christian responses to suffering and evil), 126-34, 188; and wholeness, 112-21, 174-75, 199
Health care resource allocation, xii-xiii, 206-8; and Boorse's biostatistical theory, xii, 206; Daniels's theory, xii-xiii, 206-8; and fact-value dichotomy, 206-7; Ramsey on, 188; and Rawls's theory of justice, xii, 207-8
Health-related quality of life (HRQoL), xiii, 84-85, 208
Hempel, Carl, 41n
Hick, John, 130
Ho, Justin, 8n, 52n, 75n, 80n, 167n
Hofmann, Bjørn, 25n
Holistic theories of health (Nordenfelt), 10-16, 18, 32-33, 49; concept of basic vital goals, 12, 13-14, 16, 46-47, 170; contrast to Boorse's biostatistical model, 13, 15; criticisms, 13-16; on disease/disorder, 13, 15; fact-value distinction (and value-free/value-laden), 15-16; and Fulford's reverse theory, 16-18; health and second-order abilities, 11-12; the "standard circumstances" proviso, 11-13, 14-15
Hope, Tony, 17n

Human flourishing, 174-84, 202; and Aquinas's ultimate good (the "beatific vision" or happiness), 148-49; Barth's definition of health as "strength for human life," 49, 136, 138-39, 155, 175-79, 184, 202, 207; Barth's theologically shaped understanding of, 105-6, 136-41, 175-79; Bonhoeffer's term, 12n; conceptualizing health as capacity to realize, 49; and health as a penultimate good, xvii, 140-41, 160, 181-82, 202; and health as a social and political matter, 182-84, 202; insights from disability perspectives, 99, 163-64, 178-79; insights from recent theologies of disability, 156-60; insights from the philosophical discussion, 49; and natural functions (evolutionary/biological), 179-81, 202; and radical extension of human life, 181-82, 204, 205-6; Stoics' account of, 94; tensions between subjective/objective conceptions of, 14, 170; and "the will to be healthy," 136-37, 176, 182-84, 186, 197, 202, 209; well-being and wholeness, 174-75, 199, 202

Humans as creatures, xvi, 164-74, 201-2; and Barth's account of health, 164-65, 166; and Barth's theological anthropology, 164-66, 171-74; differentiation between theological and scientific anthropologies, 166-68; particular vocational goals and particular contexts of human lives, 172-74, 201-2; penultimate or proximate ends, 170-72, 181, 201; and Thomistic/Aristotelian teleology, 165-74; and the ultimate end as union with God, 168-70, 181, 201

Humber, James M., 7n
Hume, David, 123-24n, 127-28
Humphrey, R., 66n
Hunsinger, George W., 103n
Hunt, Paul, 58
Hurst, Rachel, 69n, 72n, 73n, 74, 77n
Hyman, M., 67n

Ignatius of Antioch, 114
Impairment/disability distinctions, 56, 59-60, 61, 63-65, 189-90

Independent living movement, 58
International Classification of Functioning, Disability, and Health (ICF), xv, 71-78, 83-84, 190; as "biopsychosocial model" of disability, 72-73; concept of normal, 73-74, 190; consultation with disabled people/disability organizations, 72; definition of impairment, 73-74; definition of well-being, 83; functioning measures (Activities and Participation), 72, 74; interaction between health conditions and contextual factors, 71-73; and limits of concepts of health and disability, 83-84; Nordenfelt's objections, 74-76, 83-84; performance and capacity qualifiers, 74-75; the "present environment" stipulation, 75-76; QoL assumptions, 74
International Classification of Impairments, Disabilities and Handicaps (ICIDH), 67-71, 79; concept of normal, 69-70, 73, 79; criticisms, 68-71; medicalization of disability, 69; problems with conceptual scheme and categories for social research, 71; QoL assumptions, 70-71; use of term "handicap," 71
Iozzio, Mary Jo, 152n
Irenaeus of Lyons, 130
Iwakuma, Miho, 87

Jaspers, Karl, 167n
Jefferys, M., 67n
Jesus ben Sirach, 124, 197
Job, Book of, 129
John, Gospel of, 129-30, 131, 187
John Paul II, Pope, 155n, 193; *Salvifici Doloris*, 193

Kant, Immanuel, 90, 94n, 145-46
Kaplan, A., 71n
Karssing, E. D., 24n
Kass, Leon, 125n
Keller, Helen, 94n, 95
Khushf, George, 7-8, 12n, 13n, 15, 188n
King Asa, story of, 121, 124, 197-98
Kittay, Eva Feder, 92n

Kostanjsek, N., 67n
Kovacs, Jozsef, 12-13, 14-15, 32-33, 47n
Kripke, Saul, 95

Lammers, Stephen E., 5n, 118, 135
Latin American Pentecostalism, 111n, 116
Law, Iain, 3n, 11n, 13n; capability approach to health, 14, 15, 23-26, 89, 176-77, 184
Leder, Drew, 23, 125n
Leproi, story of the ten, 113, 119
Lilienfeld, Scott O.: prototype model (Roschian concept of disease), 36-38, 185; and Wakefield's evolutionary model of harmful dysfunction, 33-36, 81n

MacIntyre, Alasdair, 10, 95n
MacKenzie, Catriona, 85n, 86, 88-89
Marino, Lori: prototype model (Roschian concept of disease), 36-38, 185; and Wakefield's evolutionary model of harmful dysfunction, 33-36, 81n
Marmot, Michael, 182n
Mathey, Jacques, 110n, 114n
McEwan, Ian, 189n
McFadyen, Alistair, 187-89
McFarland, Ian A., 129n
McKenny, Gerald P., 23, 125
McPherson, Kath, 75n
Medical model: Barth's affirmation of doctors' roles, 138; Boorse's medical practice/medical theory distinction, 6-10; and Christian healing ministries, 115-26, 197-99, 203; Evans's theological critique of, 115-16, 118n, 119-21; and "God of the gaps" view, 123-24; phenomenological perspectives on the medical-clinical encounter, 21-23; and social model of disability, 56, 59-62; and supernatural/miraculous healings, 123, 123-24n. *See also* Socio-medical models of disability
Megone, Christopher, 9n, 31n, 118n; and Aristotle's function argument, 40-43, 47-48, 98-99; dispute with Wakefield over nature of teleological explanation, 32-33, 41-48, 168, 171, 180; on illness/disease as failures of function, 40-41; natural kinds concept, 39-40, 43, 46-48, 93; teleological account of health as natural function, 10, 38-48, 92-93, 137, 191
Meilaender, Gilbert, 124n
Mental disorders: and evolutionary models of mental functions, 27, 28n, 33-36; Fulford's bridge theory and mental illness, 17; Lilienfeld and Marino's Roschian model, 36-38, 185; "pure value" concept of, 37-38; Szasz's skeptical stance, 4, 9, 27. *See also* Cognitive impairments and intellectual disabilities; "Harmful dysfunction" model of disease and disorder (Wakefield)
Mercer, Geof, 54n, 62n, 189n
Merleau-Ponty, Maurice, 19, 21n, 87
Merriam, Garret, 93-95, 167, 172-74
Messer, Neil, 2n, 5n, 10n, 28n, 29n, 88n, 92n, 103n, 109n, 131n, 173n, 179n, 181n, 194n, 205n
Metanoia, 120
Middleton, Paul, 129n
Midgley, Mary, 210
Mieth, Dietmar, 77n
Migliore, Daniel L., 143n, 159n
Miller, Janice, 73n
Moltmann, Jürgen, 159n, 183
Morris, Jenny, 61, 65-66, 189n
Mumford, James, 156n
Murphy, Nancey, 88n

Nagi, Saad Z., 68n, 71n
National Institute for Health and Care Excellence (NICE), xiii
Natural functions: as "black box essentialist" concept, 30-31, 101n; Boorse's biostatistical model, 8-9, 79-80; evolutionary basis of, 8-9, 28-38, 41-43, 49, 79-82, 179-80; health care resource allocation and non-normative account of, 206-7; and human flourishing, 179-81, 202; and Megone's teleological account of health, 10, 38-48, 92-93, 137, 191; Wakefield-Megone dis-

pute over relationship between the good and, 41-48, 168, 171, 180. *See also* Evolutionary models of health, disease, and illness

Natural kinds, 93-98; and disability, 93-94, 95-98, 167; and Megone's Aristotelian teleological account of health as natural functions, 39-40, 43, 46-48, 93-94, 167; and species-essentialism/species-normality assumptions, 94, 97, 167. *See also* Teleological accounts of health

Nazi ideologies of health, 14-15

Nichols, Terence L., 123-24n

Nordenfelt, Lennart: concept of ability, 83-84; concept of basic vital goals, 12, 13-14, 16, 46-47, 170; holistic theory of health, 10-16, 18, 32-33, 49; objections to socio-medical model of disability in WHO's *ICF*, 74-76, 83-84; *On the Nature of Health*, 11

Normality and disability, 79-82, 164, 183; alternative concept of "responsiveness" or "individual normality," 97, 98n; Amundson's critique of disability definitions, 69-70, 80-82, 97-98, 100, 178, 190, 206; and Boorse's biostatistical model, 80; concept of normal function, 80-82, 178, 180, 206; ICF model, 79-80; ICIDH model, 79; Reynolds on the Church's "cult of normalcy," 152-54, 158; species-normality assumptions, 97-98, 177-78; Wakefield's "harmful dysfunction" model of disease and disorder, 81-82, 180, 185

Northcott, Michael, 131n, 194n

Nullard, J. B., 67n

Nussbaum, Martha: capabilities approach, xii, 14, 23, 26, 89-92, 176-79; critique of Sen's capabilities approach, 2, 26, 177; on disability and capabilities, 89-92, 178-79; *Frontiers of Justice*, 89; on Rawls's social-contract theory, 90; on Stoics' views of human capacities, 94n; ten basic capabilities, 90-91, 178

Okawa, Y., 73

Oliver, Michael, 52n; and social model of disability, 54n, 56-59, 61-65, 78-79, 82; *Understanding Disability*, 78

Oliver, Simon, 143n, 144n, 146

Otto, Rudolf, 167n

Paddison, Angus, 109n

Parens, Erik, 69n, 177n, 206n

Parish, Susan L., 57-58, 66n

Patrick, D., 67n

Paul's experience of a "thorn in the flesh" (2 Cor. 12:1-10), 118, 131, 134, 153n, 192-93

Peach, H., 67n

Peacocke, Arthur, 195n

Pentecostalism and Christian healing practices, 108n, 110, 111, 116, 121-22

Penultimate good, health as, xvii, 140-41, 160, 181-82, 202

Perry, John, 108n, 110n, 111n, 112n, 122n, 123n, 130n

Peters, Ted, xii

Peterson, Gregory R., 34n

Pfeiffer, David, 62, 68n, 69, 70, 76n, 84-85, 209n

Phenomenological perspectives on illness, 18-23, 87-89; and disability, 87-89; and disease/illness distinction, 20, 191; four levels in changing apprehension of the body wrought by illness, 20-23; Husserl's distinction between psychological and transcendental phenomenology, 19n; Iwakuma on experience of impairment in terms of the "*Umwelt*," 87; limitations and cautions, 23, 88-89; and lived experience of illness, 18-23, 87-89; the medical-clinical encounter and objectification of the body, 21-23; Merleau-Ponty, 19, 21n, 87; and positive healing relationships, 22-23, 88, 117, 199-200; Sartre on four levels of illness/pain, 19-22; and solipsism, 88; Toombs's psychological phenomenological perspective, 18-23, 50, 87, 191

Phillips, D. Z., 130-31

Philosophical accounts of health, disease,

and illness, xiv, 1-50, 163; Boorse's biostatistical model, xii, 6-10, 13, 15, 16-18, 94, 121n; capability approaches, 14, 15, 23-26, 89, 176-77, 184; critical insights for a theological account of health, 49-50; disease/illness distinctions, 1n, 6-7, 15-16, 20, 49-50, 118-19, 191; fact-value distinctions, 10, 15-16, 18, 31-33, 37-38, 50, 118-19, 180; Fulford's bridge theory, 10-11, 16-18; Lilienfeld and Marino's prototype model (disorder as Roschian concept), 36-38, 185; Megone's teleological account of health as natural function, 38-48, 92-93, 137, 191; natural kinds concept, 39-40, 43, 46-48, 93-94; Nordenfelt's holistic theory, 10-16, 18, 32-33, 49; practical orientations, 50; reverse theories (intermediate positions), 10-18; tensions between subjective/objective conceptions of human flourishing, 14, 170; terminology, 1n; Toombs's psychological phenomenological perspective, 18-23, 50, 87, 191; value-laden language, 2, 9-10, 15, 18, 32, 180; Wakefield's "harmful dysfunction" model of disease and disorder, xiv, 26-36, 100, 179-80, 185; WHO definition of health, xiv, 2-6, 62, 174

Phronēsis, 94-95
Pilch, John J., 118-19
Pinder, Ruth, 62n
Placher, William C., 103n
Pope, Andrew M., 68n, 71n
Porter, Jean, 95, 173
Portmann, Adolf, 167n
Prenatal diagnosis and abortion, 85-86, 155-56n
Price, Janet, 64n, 70n
Prudence *(prudentia)*, 95, 172-74, 185-86
Putnam, Daniel, 52n

Quality of life (QoL), xiii, 84-86, 208-9; assumptions about disability, 70-71, 74, 84-86, 208-9; health-related (HRQoL), xiii, 84-85, 208; and sociomedical models of disability, 70-71, 74; WHO's *ICF* model, 74; WHO's *ICIDH* model, 70-71
Quality-adjusted life year (QALY), xiii, 85, 208

Radical life extension, 181-82, 194n, 204, 205-6
Ralston, D. C., 8n, 52n, 75n, 80n, 167n
Ramsey, Paul, 188
Rationality: and disability perspectives, 90, 93, 169; and teleological accounts of health, xvi, 40-42, 98-99, 149-50, 169
Rawls, John, xii, 90, 207-8
Reed, John Shelton, 5n
Reformed tradition, 105-6
Rehmann-Sutter, Christoph, 77n
Reinders, Hans S., 14n, 154, 155-57; reading the *imago Dei* in terms of communal relations, 156-57; *Receiving the Gift of Friendship*, 155-56; on rights-based project of inclusion, 154
Reiss, Michael J., 194n
"Reverse theories" of health, disease, and illness, 10-18
Reynolds, Thomas E., 14n, 152-55, 158, 159n
Riddell, Sheila, 100n
Ridley, Mark, 32n
Rogers, Eugene F., Jr., 142, 165-66
Romero, Miguel J., 150n
Rosch, Eleanor R., 36n
Roschian understanding of mental disorder, 36-38, 185
Rose, Steven, 36n
Ruse, Michael, 47n
Rutherford, Adam, 117-18, 121, 123

Sacks, Oliver, 22n, 199-200
Salek, Sam, 85n
šālôm (shalom), 112, 174
Sartre, Jean-Paul, 19-22
Schickle, Darren, 85n
Schloss, Jeffrey, 181n
Schneider, M., 67n
Schneidert, Marguerite, 73
Schramme, Thomas, 11n, 13-14
Scott, Peter Manley, 141n, 182n

Scully, Jackie Leach, 85n, 86, 88-89, 100n
Sedgwick, Peter, 37
Seedhouse, David, 12n
Seelman, Katherine D., 23n, 52n, 179n, 209n
Sen, Amartya, xii, 2, 14, 23-26, 89, 176-77
Shakespeare, Tom, 52n, 58, 61, 63-64, 77, 86n, 190
Shelton, James B., 108n, 119n
Shildrick, Margrit, 64n, 70n
Shuman, Joel, 197n
Silvers, Anita, 52n; criticism of concept of the normal, 92, 177-78, 206; and social models and theories of disability, 54, 69n, 70n, 77-78, 79n, 92
Simmons, Fred, 194n
Sin: and Aquinas's teleological understanding of human life, 148; complex connections between disease and, 130n, 186-89, 191-93, 203; forgiveness of, 114, 130n; as idolatry, 187; McFadyen on theological language of, 187-89. *See also* Evil, suffering, and sin (disease and)
Slave societies and ideas of health, 14-15
Slovo, Gillian, 5n
Smith, C., 67n
Social model of disability, 54-78, 99, 189-90; and cognitive impairments, 66; contrasted with individual model, 56-57; critiques by disabled feminists, 65-66; differences between the two models, 58-59; and disability rights movement, 55-59, 63, 66; impairment/disability distinction, 56, 59-60, 61, 63-65, 189-90; impairment/disease distinction, 62-63, 78-79, 189-90; issue of coverage and representation, 65-66; medical intervention question, 61-62, 189-90; and medical model/medicalization of disability, 56, 59-62; "minority-group" model (U.S.), 55, 57-59; recent criticisms and clarifications, 59-66; and scope/limits of health and disability, 82-83; social model (British), 54, 55-66, 99, 189-90; terminology ("disabled people" or "people with disabilities"), 52-53; and watershed civil rights legislation (U.S.), 57-58; and WHO definition of impairment, 62; and WHO's *ICIDH* definitions, 68-71

Socio-medical models of disability, 54, 67-78; concept of the normal, 69-70, 73-74, 79, 190; conceptual schemes and categories for social research, 71; consultations with disabled people/disability organizations, 69, 72; criticisms by disability activists and social-model authors, 68-71; definition of impairment, 73-74; definition of well-being, 83; functioning measures, 72, 74; interaction between health conditions and contextual factors, 71-73; and limits of concepts of health and disability, 83-84; medicalization of disability, 69; Nordenfelt's objections, 74-76, 83-84; performance and capacity qualifiers, 74-75; the "present environment" stipulation, 75-76; QoL assumptions, 70-71, 74; term "handicap," 71, 76; WHO's *ICF* (2001 revision), xv, 71-78, 83-84, 190; WHO's *ICIDH* definitions, 67-71, 79
Solipsism, 88
Song, Robert, xii
Souls and bodies (Christian theological anthropology), 88n
Sōzō, 113
Sparrow, Robert, 100n
Species-essentialism, 94, 97, 167
Stahl, Elisabeth, xiii
Stoics, 94
"Strategies for engineered negligible senescence" (SENS), 182n, 194n. *See also* Radical life extension
Suffering. *See* Evil, suffering, and sin (disease and)
Sulmasy, Daniel P., 95-98, 167
Swain, John, 52n
Swinton, John, 131, 150n, 158n, 178
Szasz, Thomas S., 4, 6, 9, 27
Szerszynski, Bronislaw, 173n

Tarlov, Alvin R., 68n, 71n
Teleological accounts of health, 38-48, 92-99, 141-50; Aristotle's function ar-

gument, 40-43, 47-48, 98-99; and Barth's ethics of creation, 141-44, 165-66; the body and Thomistic teleology, 148-49, 169; and disability, 92-99, 101; distinctions between intrinsic and extrinsic teleologies, 145-46; human beings as substances, 144-45; and human rationality, xvi, 40-42, 98-99, 149-50, 169; Megone on illness and disease as failures of function, 40-41; Megone's account of natural functions, 38-48, 92-93, 137, 191; Merriam's exploration of Aristotelian virtue ethics, 93-95, 167, 172-74; notion of natural kinds, 39-40, 43, 46-48, 93-98, 167; and practical wisdom *(phronēsis)*, 94-95, 172-74; and prudence *(prudentia)*, 95, 172-74, 185-86; species-essentialism or species-normality assumptions, 94, 97, 167; Thomistic teleology, xvi, 141-50, 155, 165-74; and the ultimate good (Thomistic "beatific vision" or happiness), 147-49, 168-69; Wakefield/Megone dispute over nature of teleological explanation, 41-48, 168, 171, 180; Wakefield's "harmful dysfunction" model of disease and disorder, 29, 41-48, 146, 168, 171, 180

Theological literature on disability, 14n, 151-60, 169; and disability rights discourse, 154-57; Eisland's liberation theology of disability, 152-54; eschatological perspectives, 160, 169; Hauerwas and Willimon, 151-52, 157, 174, 186, 205; Hauerwas on practices of Christian communities, xvii, 151-52, 186, 199-200; and human flourishing (what constitutes), 156-60; image of Jesus as disabled God, 153-54; reading the *imago Dei* in terms of communal relations, 156-57; reflections on intellectual disabilities, 151-60, 169, 186, 205; Reinders, 154, 155-57; Reynolds on the Church's "cult of normalcy," 152-54, 158; Swinton on participation in worship and Eucharist, 159, 169; Volpe, 154, 155; Yong, 154-55, 157, 160, 169

Theological resources for understanding health and disease, xv-xvi, 103-61, 164; Barth's ethics of creation, xv-xvi, 134-41, 164-66, 171-74; Christian healing ministries and care for the sick, xv, 107-34; Christian responses to suffering and evil (theodicy), 126-34, 188; health as wholeness, 112-21, 174-75, 199; recent theologies of disability, 14n, 151-60, 169; relationship between external influences and internal sources (biblical texts), 103-4; Thomistic teleology, xvi, 141-50, 155, 165-74

Theological theses concerning health, disease, and illness, xvi-xvii, 163-200; disease, suffering, evil, and sin, 130n, 184-97, 202-3; health and creaturely flourishing, 174-84, 202; humans as creatures, xvi, 164-74, 201-2; practical implications, 197-200, 203

Thomas, Carol, 66

Thomas Aquinas: account of evil as failure to realize proper end, 150; and Aristotle, 144-47, 165-66; attempted *rapprochements* between Barth and, 141-44, 165-66; and *prudentia*, 95, 172-74; teleological understanding of human life, xvi, 141-50, 155, 165-74; and the ultimate good (the "beatific vision" or happiness), 147-49, 168-69. *See also* Teleological accounts of health

Thrall, Margaret E., 135n

A Time to Heal (Church of England report), 109, 110, 112, 116-17, 122-23, 130-31

Titus, Arthur, 167n

Toombs, S. Kay: disease/illness distinction, 20, 191; *The Meaning of Illness*, 18; on more positive healing relationships, 22-23, 88, 117, 199-200; psychological phenomenological perspective on illness, 18-23, 50, 87, 191

Topliss, Eda, 52n

Torrance, Thomas F., 25n

Tremain, Shelley, 63-64, 99

Trivers, Robert, 28n

Ueda, S., 73

Uexküll, Jakob von, 87n

Union of the Physically Impaired Against Segregation (UPIAS), 55-56, 58-59, 63, 66
Üstün, T. B., 67n, 72n, 73n

Vácha, Jiří, 97
Vaux, Kenneth, 176n
Veatch, Robert, 115
Verhey, Allen, 5n, 118, 135
Verkerk, M. A., 24n
Volck, Brian, 197n
Volpe, Medi Ann, 66n, 154, 155

Wakefield, Jerome C.: and Aristotelian teleology, 29, 41-48, 146, 168, 171, 180; dispute with Megone over nature of teleological explanation (and relationship between natural function and the good), 41-48, 168, 171, 180; and evolutionary basis of natural functions, 8-9, 28-38, 41-43, 49, 81-82, 179-80; and exaptations, 33-35; "harmful dysfunction" model, xiv, 26-36, 100, 179-80, 185; response to critics, 34-35, 38n. *See also* "Harmful dysfunction" model of disease and disorder (Wakefield)
Walters, LeRoy, 14n
Wannenwetsch, Bernd, 195, 196
Warren, M. D., 67n
Warrington, Keith, 108n, 111, 121-22, 124
Wasserman, David, 52, 56, 67, 82n
Waters, Brent, 141n, 182n, 205n
Watson, Nicholas, 61n, 100n
Well-being, 1n; and socio-medical models of disability, 83; wholeness and human flourishing, 174-75, 199, 202; WHO's definition of health, 3-4, 83, 174; WHO's *ICF* definition, 83
Wells, Samuel, 133n, 197n
Wholeness: Barth's notion of health as, 138, 175, 176; Christian healing ministries and emphasis on health as, 112-21, 174-75, 199; and critiques of the medical model, 115-21; and distinction between curing disease and healing illness, 116-21; Evans's argument for theological understanding of health as, 115-16, 118n, 119-21; and Hebrew word *šālôm*, 112, 174; and human flourishing, 174-75, 199, 202; and Nordenfelt's holistic theory of health, 10-16, 18, 32-33, 49
Widdows, Heather, 3n, 11n, 13n; capability approach to health, 14, 15, 23-26, 89, 176-77, 184
Wiggins, David, 95
Wilkinson, John, 6n, 112, 174n
Wilkinson, Richard G., 182n
Williams, Alan, 85n
Williams, Gareth, 23, 77, 87-88
Willimon, William H., 151-52, 157, 174, 186, 200, 205
Wilson, David Sloan, 33n
Wolff, Christian, 144n
World Council of Churches' Christian Medical Commission, definition of health, 112-13
World Health Organization (WHO), definition of health, xiv, 2-6, 62, 83, 112, 174, 183; all-embracing character (medicalization of life), 4-6, 15, 83, 118; contrasted to Boorse's biostatistical model, 6-10; and definition of impairment, 60, 62; impairment/disability distinction, 60; problems/drawbacks, 3-6; and spiritual well-being, 5-6; term "well-being," 3-4, 83, 174; vagueness, 3-4
World Health Organization (WHO), socio-medical models of disability: *ICF*, xv, 71-78, 83-84, 190; *ICIDH* definitions, 67-71, 79

Yates, Anne, 17n
Yong, Amos, 84n, 154-55, 157, 160, 169

Zalta, Edward N., 52n
Zaner, Richard M., 20-21, 23, 125n
Zarb, Gerry, 65n
Zöckler, Otto, 167n